CARDIAC ENERGETICS:
FROM Emax TO PRESSURE-VOLUME AREA

DEVELOPMENTS IN CARDIOVASCULAR MEDICINE

121. S. Sideman, R. Beyar and A. G. Kleber (eds.): Cardiac Electrophysiology, Circulation, and Transport. Proceedings of the 7th Henry Goldberg Workshop (Berne, Switzerland, 1990). 1991. ISBN 0-7923-1145-0.
122. D. M. Bers: Excitation-Contraction Coupling and Cardiac Contractile Force. 1991. ISBN 0-7923-1186-8.
123. A.-M. Salmasi and A. N. Nicolaides (eds.): Occult Atherosclerotic Disease. Diagnosis, Assessmentand Management. 1991. ISBN 0-7923-1188-4.
124. J. A. E. Spaan: Coronary Blood Flow. Mechanics, Distribution, and Control. 1991. ISBN 0-7923-1210-4.
125. R. W. Stout (ed.): Diabetes and Atherosclerosis. 1991. ISBN 0-7923-1310-0.
126. A. G. Herman (ed.): Antithrombotics. Pathophysiological Rationale for Pharmacological Interventions. 1991. ISBN 0-7923-1413-1.
127. N. H. J. Pijls: Maximal Myocardial Perfusion as a Measure of the Functional Significance of Coronary Arteriogram. From a Pathoanatomic to a Pathophysiologic Interpretation of the Coronary Arteriogram. 1991. ISBN 0-7923-1430-1.
128. J. H. C. Reiber and E. E. v. d. Wall (eds.): Cardiovascular Nuclear Medicine and MRI. Quantitation and Clinical Applications. 1992. ISBN 0-7923-1467-0.
129. E. Andries, P. Brugada and R. Stroobrandt (eds.): How to Face "the Faces" of Cardiac Pacing. 1992. ISBN 0-7923-1528-6.
130. M. Nagano, S. Mochizuki and N. S. Dhalla (eds.): Cardiovascular Disease in Diabetes. 1992. ISBN 0-7923-1554-5.
131. P. W. Serruys, B. H. Strauss and S. B. King III (eds.): Restenosis after Intervention with New Mechanical Devices. 1992. ISBN 0-7923-1555-3.
132. P. J. Walter (ed.): Quality of Life after Open Heart Surgery. 1992. ISBN 0-7923-1580-4.
133. E. E. van der Wall, H. Sochor, A. Righetti and M. G. Niemeyer (eds.): What is new in Cardiac Imaging? SPECT, PET and MRI. 1992. ISBN 0-7923-1615-0.
134. P. Hanrath, R. Uebis and W. Krebs (eds.); Cardiovascular Imaging by Ultrasound. 1992. ISBN 0-7923-1755-6.
135. F. H. Messerli (ed.): Cardiovascular Disease in the Elderly, 3rd ed. 1992. ISBN 0-7923-1859-5.
136. J. Hess and G. R. Sutherland (eds.); Congenital Heart Disease in Adolescents and Adults. 1992. ISBN 0-7923-1862-5.
137. J. H. C. Reiber and P. W. Serruys (eds.): Advances in Quantitative Coronary Arteriography. 1993. ISBN 0-7923-1863-3.
138. A.-M. Salmasi and A. S. Iskandrian (eds.): Cardiac Output and Regional Flow in Health and Disease. 1993. ISBN 0-7923-1911-7.
139. J. H. Kingma, N. M. van Hemel and K. I. Lie (eds.): Atrial Fibrillation, a Treatable Disease? 1992. ISBN 0-7923-2008-5.
140. B. Ostadal, N. S. Dhalla (eds.): Heart Function in Health and Disease. 1993. ISBN 0-7923-2052-2.
141. D. Noble and Y.E. Earm (eds.): *Ionic Channels and Effect of Taurine on the Heart.* Proceedings of an International Symposium (Seoul, Korea, 1992). 1993. ISBN 0-7923-2199-5.
142. H.M. Piper and C.J. Preusse (eds.): *Ischemia-reperfusion in Cardiac Surgery.* 1993. ISBN 0-7923-2241-X.
143. J. Roelandt, E.J. Gussenhoven and N. Bom (eds.): *Intravascular Ultrasound.* 1993. ISBN 0-7923-2301-7.
144. M.E. Safar and M.F. O'Rourke (eds.): *The Arterial System in Hypertension.* 1993. ISBN 0-7923-2343-2.
145. P. W. Serruys, D.P. Foley and P.J. de Feyter (eds.): *Quantitative Coronary Angiography in Clinical Practice.* 1993. ISBN 0-7923-2368-8.
146. J. Candell-Riera and D. Ortega-Alcalde (eds.): *Nuclear Cardiology in Everyday Practice.* 1993. ISBN 0-7923-2374-2.
147. P. Cummins (ed.): *Growth Factors and the Cardiovascular System.* 1993. ISBN 7923-2401-3.
148. K. Przyklenk, R.A. Kloner and D.M. Yellon (eds.): *Ischemic Preconditioning: The Concept of Endogenous Cardioprotection.* 1993. ISBN 0-7923-2410-2.

CARDIAC ENERGETICS:
FROM Emax TO PRESSURE-VOLUME AREA

edited by

Martin M. LeWinter
Cardiology Unit
Department of Medicine
University of Vermont
Burlington, Vermont, USA

Hiroyuki Suga
Department of Physiology
Okayama University Medical School
Okayama, Japan

Matthew W. Watkins
Cardiology Unit
Department of Medicine
University of Vermont
Burlington, Vermont, USA

Kluwer Academic Publishers
Boston/London/Dordrecht

Distributors for North America:
Kluwer Academic Publishers
101 Philip Drive
Assinippi Park
Norwell, Massachusetts 02061 USA

Distributors for all other countries:
Kluwer Academic Publishers Group
Distribution Centre
Post Office Box 322
3300 AH Dordrecht, THE NETHERLANDS

Library of Congress Cataloging-in-Publication Data

A C.I.P. Catalogue record for this book is available
from the Library of Congress.

Printed on acid-free paper.

Printed in the United States of America

Contents

List of Contributors

Hidetsugu Asanoi, MD, PhD
Second Department of Medicine
Pharmaceutical University
Toyama, JAPAN

Co-authors: Tomoki Kameyama and Shinji Ishizaka

Daniel Burkhoff, MD
Assistant Professor of Medicine
Division of Circulatory Physiology
Columbia Presbyterian Hospital
177 Fort Washington Avenue
New York, NY 10032

Co-authors: Matthew Schnellbaecher, Richard Stennett, Donna Zwas, Kazuhide Ogino and James Morgan

Masao Endoh, MD
Professor
Department of Pharmacology
Yamagata University
School of Medicine, 2-2-2
Iida-nishi, 990-23
Yamagata, JAPAN

Gregory Freeman, MD
Professor of Medicine and Physiology
Division of Cardiology
Department of Medicine
The University of Texas
Health Science Center at
San Antonio
San Antonio, TX 78284-7872

Co-author: Sumanth Prabhu

Yoichi Goto, MD
Division of Cardiology
Department of Medicine
National Cardiovascular Center, Suita
Osaka, JAPAN

Co-authors: Satoshi Yasuda, Toshiaki Shishido, Kengo Fukushima and Shogo Suzuki

Fumihiko Kajiya, MD, PhD
Professor, Department of Biomedical Engineering and Systems Cardiology
Kawasaki Medical School
Matsushima, Kurashiki
701-01, JAPAN

Co-authors: Fumihiko Kajiya, Toyotaka Yada, Yasuo Ogasawara, Shigeru Ohta, Osamu Hiramatsu, Masami Goto, and Katsuhiko Tsujioka

Hideo Kusuoka, MD, PhD
Associate Professor
Biomedical Research Center
Osaka University Medical School
Suita, Osaka 565
JAPAN

Co-author: Eduardo Marban

Martin LeWinter, MD
Professor of Medicine
University of Vermont
McClure 1 - MCHV
Burlington, VT 05401

Co-authors: Akihiro Higashiyama, Hitoshi Yaku and Matthew Watkins

John Solaro, MD
Professor and Chair
Department of Physiology and Biophysics
University of Illinois
901 South Wolcott Avenue
Chicago, IL 60612-7342

Hiroyuki Suga, MD
Professor and Chairman
Department of Physiology II
Okayama University Medical School
2-5-1 Shikatacho, Okayama 700
JAPAN

Co-authors: Miyako Takaki, Hiromi Matsubara and Yoichi Goto

Motoaki Sugawara, PhD
Professor, Department of Cardiovascular Sciences
The Heart Institute of Japan
Tokyo Women's Medical College
8-1 Kawada-cho, Shinjuku-ku
Tokyo, 162 JAPAN

Co-authors: Yukiyoshi Kondoh and Kiyoharu Nakano

Seiryo Sugiura, MD
The Second Department of Internal Medicine
Tokyo University School of Medicine
7-3-1 Hongo, Bunkyo-ku
Tokyo 113, JAPAN

Co-authors: Hiroshi Yamashita, Masataka Sata, Hideo Fujita, Shin-ichi Momomura, Takashi Serizawa, and Haruo Sugi

Kenji Sunagawa, MD, PhD
Director, Department of Cardiovascular Dynamics
National Cardiovascular Center Research Institute
5-7-1 Fujishiro-dai, Suita
Osaka, JAPAN

Co-authors: Masaru Sugimachi, Yasuhiro Ikeda, Osamu Kawaguchi, Toshiake Shishido and Toru Kawada

Motishi Takeuchi, MD
The First Department of Internal Medicine
Kobe University School of Medicine
7-5-2 Kusunoki-cho Chuo-ku
Kobe, Hyogo, 650
JAPAN

Tad Taylor, MD, PhD
Resident
California Pacific Medical Center
3700 California Street
San Francisco, CA 94118

Co-authors: Yoichi Goto and Hiroyuki Suga

Mitsuhiro Yokota, MD
Cardiovascular Disease
Nagoya University Hospital
65 Tsuramai-cho, Showa-ku
Nagoya, JAPAN

Co-authors: Hitoshi Ishihara and Toshikazu Sobue

Preface

The energetics of cardiac muscle and whole heart contraction have fascinated scientists and clinicians for many years. There are undoubtedly many reasons for this. Problems are seemingly tractable because muscle contraction has a more readily quantifiable, biophysical basis compared to many other areas in biology. Thus, researchers interested in cardiovascular modeling find this a challenging, yet fruitful area. Similarly, those interested in structure-function relationships have been intrigued by the links between the mechanical performance of molecular "motors" and energy consumption. Cardiac energetics is also an area in which it has been increasingly possible to relate discoveries at a very basic level to the performance of the whole, integrated organ. Finally, the clinical relevance of alterations in energy utilization in cardiac hypertrophy and failure, both as a "window" with which to understand basic mechanisms and as potential therapeutic targets, has been widely recognized.

As a reflection of the increasing interest in and progress made in cardiac energetics, especially over the last 10-15 years, the Japan-USA Cardiac Energetics Workshop was held at Okayama University, Okayama, Japan, in February 1994. Much of the progress made during this period of time has been spurred by the pressure-volume area concept, the natural extension into energetics of earlier, pioneering work delineating the time varying elastance framework for ventricular contraction. The title of this volume reflects this linkage. The organizers of the Workshop attempted to bring together a spectrum of researchers, basic, applied and clinical, with a shared interest in the energetics of cardiac muscle and ventricle, in order to provide an overview of the current "state of the art". The Workshop was lively, informative and provocative. We hope that this volume, which includes chapters by virtually all of the participants, reflects the excitement of the Workshop and will be valuable to scientists and clinicians alike. The editors are indebted to the Workshop participants and their colleagues for making this volume possible. In addition, the support of the Suzuken Memorial Foundation, Nagoya, Japan, the

Technopolis Foundation of Okayama Prefecture, Japan, and Merck, Inc., USA, are gratefully acknowledged. Finally, the secretarial assistance of Beverly Cooper and editorial guidance and patience of Melissa Welch of Kluwer Academic were absolutely essential in its preparation.

Martin M. LeWinter, MD
Cardiology Unit
Department of Medicine
University of Vermont
Burlington, Vermont, USA

Hiroyuki Suga, MD
Department of Physiology II
Okayama University
Okayama, Japan

Matthew W. Watkins, MD
Cardiology Unit
Department of Medicine
University of Vermont
Burlington, Vermont, USA

1. Energy Costs of PVA and Emax:Constancy and Variability

Hiroyuki Suga, Miyako Takaki, Hiromi Matsubara
and Yoichi Goto

INTRODUCTION

For improving the present methods of diagnosis, treatment and prevention of cardiac diseases, a much better understanding of cardiac mechanoenergetics, or contractile performance and energy demand, is mandatory. We have devoted ourselves to discovering a ventricular contractility index which is as much load independent as possible and to elucidating primary determinants of cardiac oxygen consumption with a sound physiological basis.

As the results of our energetic efforts together with good luck, we have found that left ventricular oxygen consumption (Vo_2) could reasonably be formulated as a function of a ventricular contractility index (Emax: end-systolic maximal elastance or pressure-volume ratio, basically interchangeable with Ees) and a measure of ventricular total mechanical energy (PVA: systolic pressure-volume area). These concepts of Emax and PVA were first proposed by the first author (Suga) nearly 30 and 20 years ago, respectively (1, 2). He made great efforts to establish the concept of Emax with the late Dr. Kiichi Sagawa, a Johns Hopkins University Professor, from 1970-1978 (3,4), and also to establish the concept of PVA with many excellent energetic collaborators in Japan since 1978 (5-7).

For the discovery of these two mechanical and energetic concepts, Suga was nominated as the 1993 Paul Dudley White International Lecturer at the 66th American Heart Association (AHA) Scientific Sessions, Atlanta (8), and awarded the 1993 Upjohn Scientific Research Fellowship. He was also nominated as the 1994 Witzig Lecturer of the 11th International Conference of the Cardiovascular System Dynamics Society (CSDS), San Francisco.

This chapter briefly summarizes recent experimental and theoretical advances in cardiac mechanoenergetics based on the Emax and PVA concepts, about which our research group (at National Cardiovascular Center of Japan, 1978-1991, and Okayama University, 1991-present) is most confident.

Emax and PVA

Figure 1 schematically explains the Emax and PVA concept. Emax has been shown to quantify contractility of the ventricular chamber (1,3,4). PVA has been shown to quantify the total mechanical energy generated by ventricular contraction (2,5-7).

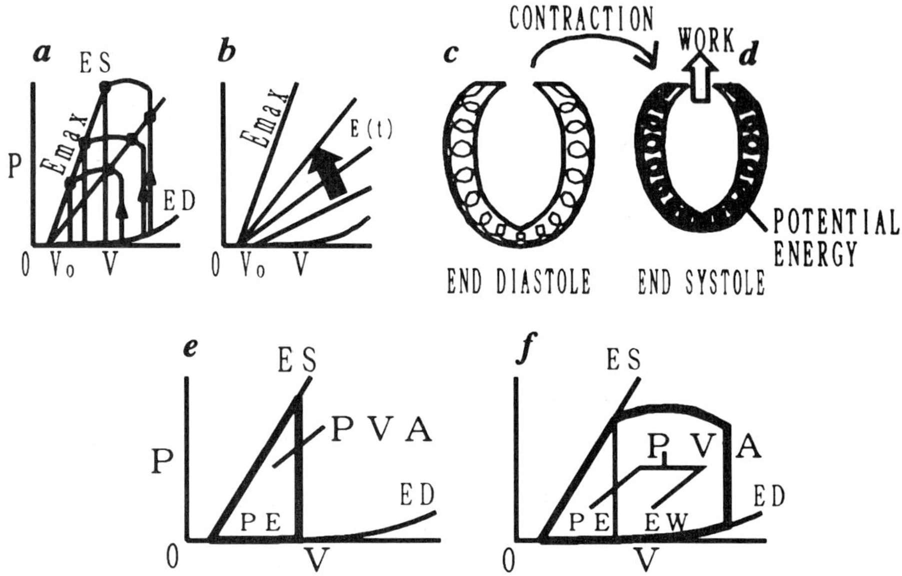

Fig 1. Schematic explanation of Emax and PVA.

The contractility index Emax was first defined as the slope of the end-systolic pressure-volume relation which was relatively linear in normal canine left ventricles (1,3,4). In Figure 1, Panel *a* is the pressure (P)-volume (V) diagram depicting three P-V loops under different preload and afterload conditions enveloped by the end-systolic (ES) and end-diastolic (ED) P-V relations. The left three circles are end-systolic P-V points connected by the end-systolic P-V line. The right three circles are P-V points at a specified intermediate time during systole, connected by an instantaneous P-V line. Vo is the volume-axis intercept of the end-systolic and instantaneous P-V lines.

Panel *b* illustrates counterclockwise rotation of the instantaneous P-V line during systole. The time-varying slope of the instantaneous P-V line is represented by the time-varying elastance E(t); E stands for volume elastance and E(t) stands for elastance that changes with time. Emax stands for end-systolic maximal elastance.

In Panel *c*, the thin spring in the wall of the cardiac chamber symbolizes the end-diastolic compliant or minimally elastic chamber wall. In Panel *d*, the thick spring symbolizes the end-systolic stiffened or maximally elastic chamber wall. The increment in elastance during systole is accompanied by a mechanical work output or external work performed against the pressure load outside of the chamber and an increment in elastic potential energy during systole within the time-varying elastic wall.

The ventricular energy measure PVA was first defined as a specific area under the end-systolic P-V relation line (2). Panel *e* shows an isovolumic contraction which is accompanied by no external work output and an increment in potential energy (PE) in the ventricular wall. Panel *f* represents an ejecting contraction which is accompanied by some external work output (EW) and an increment in potential energy (PE). The sum of EW and PE is the total mechanical energy that the time-varying elastic chamber generates during systole. PE is considered to be fully degraded into heat during relaxation (6). In physics, this energy can be quantified by the specific area (pressure-volume area: PVA) bounded by the end-systolic and end-diastolic P-V relations and the systolic P-V trajectory as shown by the area bounded by the heavy lines in Panels *e* and *f*. Although Emax and PVA are bulk variables of the ventricle, we consider that they spatially and temporally integrate mechanoenergetics (or unitary elasticity and unitary energy consumption) of all the crossbridge attachment and detachment cycles involved in ventricular pressure development during contraction (9-11).

Oxygen Costs of PVA and Emax

In mechanoenergetic studies on Emax and PVA, we have been using the left ventricle of the excised cross-circulated canine heart preparation for nearly 20 years since Suga devised it with the late Dr. Sagawa (4). Although this heart preparation is excised and cross-circulated, it has the advantage that the heart never suffers from myocardial ischemia even for a short time during the surgical preparation (4). In addition, the heart is free from the sympathetic and vagal neural influences of the cardiovascular center, but dependent on humoral factors in the coronary arterial blood coming from the support dog's systemic artery. Therefore, good maintenance of the support dog's condition including anesthetic level, systemic blood pressure, blood gases and pH is mandatory to obtain stable preparations. We have found the following mechano-energetic properties with respect to PVA and Emax (6,7).

In a stable contractile state, Vo_2 linearly correlates with PVA regardless of the mode of contraction, namely, whether the contraction is isovolumic or ejecting under various preload and afterload conditions (5-7). Figure 2 Panel *a* shows schematically the Vo_2-PVA relation in a stable contractile state. The slope (a) of the relation indicates the O_2 cost of PVA or mechanical energy. The relation has a positive Vo_2-axis intercept (b).

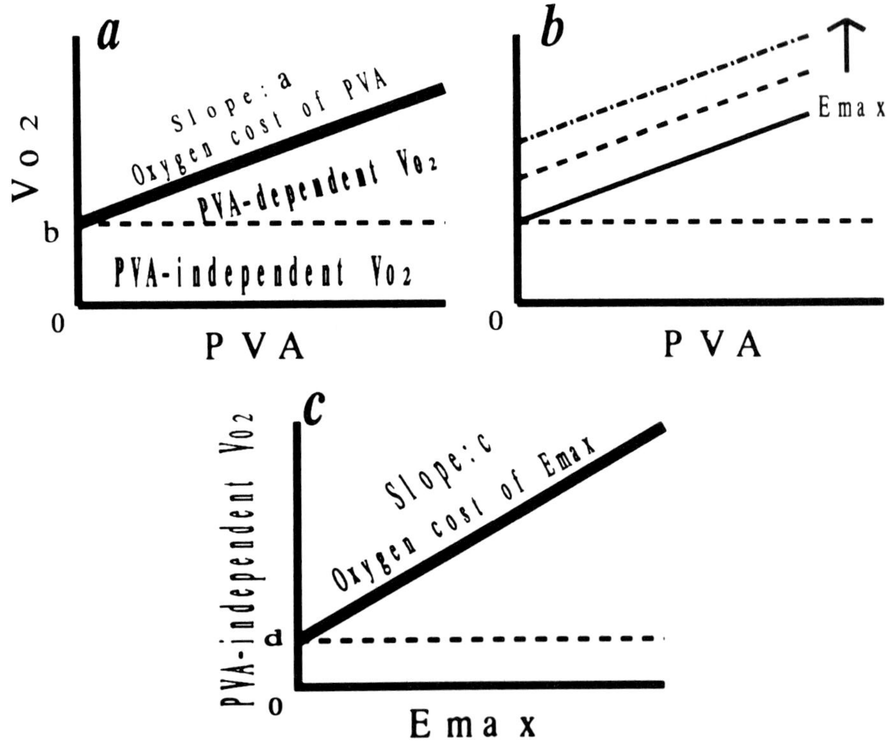

Fig. 2. Schematic illustration of the left ventricular Vo_2-PVA relationship.

Total Vo_2 can be divided into the PVA-dependent fraction above b and the PVA-independent Vo_2 equal to b. The reciprocal of the slope (1/a) indicates the contractile efficiency from PVA-dependent Vo_2 to PVA. The contractile efficiency is considered to be the product of conversion efficiency of PVA-dependent Vo_2 to ATP by mitochondrial oxidative phosphorylation and from ATP to PVA by the contractile machinery (7, 9).

A similar relation has been confirmed in papillary muscles (12) and rabbit hearts (13). The load-independent Vo_2-PVA relation accounts for the cardiac muscle Fenn effect, i.e., load-dependent energetics (14,15). These facts corroborate the physiological significance of PVA as the primary mechanical determinant of myocardial energy consumption (6, 7, 9).

PVA-independent Vo_2 is considered to be related primarily to Ca^{++} handling for excitation-contraction (E-C) coupling and basal metabolism to maintain myocardial viability and integrity (6). We consider that Na^+ handling energy for membrane excitation is relatively small (6).

The Vo_2-PVA relation shifts its elevation with increments in Emax with most positive inotropic interventions including catecholamines (epinephrine, norepinephrine, isoproterenol, dobutamine, denopamine, etc), Ca^{++}, digitalis (ouabain), paired pulse stimulation and various new cardiotonic agents (sulmazole, milrinone, OPC 8212, DPI 201-106, UDCG 115 BS, EMD 53998) (6,7,16-18, some unpublished) and negative inotropic interventions including β blockers (propranolol, nipradilol), Ca antagonists (verapamil), anesthetics (pentobarbital, fentanyl) and 2,3 butanedione monoxime (BDM)[6, 7, 19-21, some unpublished], as shown schematically in Figure 2 Panel *b*.

Despite the elevation changes, the slope (a) of the Vo_2-PVA relation, namely, O_2 cost of PVA, and its reciprocal, contractile efficiency, are largely constant under various positive and negative inotropic interventions mentioned above.

There are however some exceptions to the constancy of the O_2 cost of PVA. These are stunned (22) and acidotic hearts (23), both of which have a smaller O_2 cost of PVA. We do not yet know the mechanisms underlying the reduced O_2 cost of PVA or an enhanced contractile efficiency (22,23). For the failing heart, this change seems favorable because the same PVA can be generated by a smaller O_2 consumption, particularly when O_2 supply and hence ATP product are limited. The underlying mechanisms which cause this energetic improvement must be elucidated.

Another exception is the hyperthyroid rabbit heart which has a greater O_2 cost of PVA (24). Hyperthyroidism in associated with a myosin isozyme shift from β (lower ATPase activity, more economical) to α (higher ATPase activity, less economical) type in the rabbit heart. This phenotype change of myosin may be the cause of the increased O_2 cost of PVA and decreased contractile efficiency (24). However, hyperthyroid canine hearts have the same O_2 cost of PVA and contractile efficiency as euthyroid ones (25). These hyperthyroid canine hearts did not show any significant myosin isozyme shift to α type nor a significant increase in myosin ATPase activity although the myocardial force transient response to length perturbation was 20-70% augmented suggesting an increased crossbridge cycling rate (25).

The slope (c) of the PVA-independent Vo_2-Emax relation, which we call the O_2 cost of Emax (see Figure 2, Panel *c*), is also constant for various positive and negative inotropic interventions (6-8,16,17,20,21,26) including several new cardiotonic agents (18,27,28). We consider that changes in the PVA-independent Vo_2 are mainly due to changes in Ca^{++} handling energy for E-C coupling (6,7). We consider that changes in basal metabolism and Na^{+} handling energy are relatively small (6). Based on the tight stoichiometry of Ca^{++}:ATP coupling for Ca^{++} handling by the sarcoplasmic Ca^{++} pump ATPase (6), we assume that a change in PVA-independent Vo_2 indirectly indicates a proportional change in the total amount of Ca^{++} handled for E-C coupling.

However, cardiac cooling (29), warming (30) and vibration (31) decrease the O_2 cost of Emax, and stunning (22) and acidosis (23) increase the O_2 cost of Emax. Although ryanodine resulted in a disproportionally high PVA-independent VO_2 despite gradually decreased Emax (32), its effect on the O_2 cost of Emax remains unknown.

Thus, we have found that both the O_2 cost of PVA and Emax are usually constant for various inotropic interventions, but change variably during pathophysiological conditions (7). In addition, we are curious why a Ca sensitizer did not decrease the O_2 cost of Emax despite our expectation that this would occur.

Simulation

To get a better insight into the mechanisms underlying the constancy and variability of the O_2 costs of PVA and Emax under various inotropic interventions and pathophysiological conditions, we carried out the following computer stimulation studies. The simulation consisted of Ca cycling during excitation-contraction (E-C) coupling and CB cycling triggered by a Ca transient (33). The Ca cycling consisted of Ca release from the sarcoplasmic reticulum (SR), Ca binding to troponin C and Ca uptake (rate constant k_3) by the SR. Ca association and dissociation rate constants (k_1 and k_2) with troponin were varied. Ca sensitivity of troponin C (Ka) was given by k_1/k_2 ratio. Ca bound to troponin triggered CB cycling. CB attachment (on) and detachment (off) rate constants (f and g) were varied.

The equations we solved were:

$$d[Ca^{++}]/dt = Ca^{++} \text{ release rate} - k_3[Ca^{++}] - d[TnCa]/dt$$

$$d[TnCa]/dt = k_1[Ca^{++}][\text{total Tn} - TnCa] - k_2[TnCa]$$

$$d[onCB]/dt = f[TnCa][\text{total available CB} - onCB] - g[onCB]$$

We assumed PVA $\propto$ Emax $\propto$ Peak force $\propto$ Peak [onCB] at a given muscle length. We considered only sarcomere-isometric contractions.

Solving these equations, we obtained the results as shown in Figure 3.

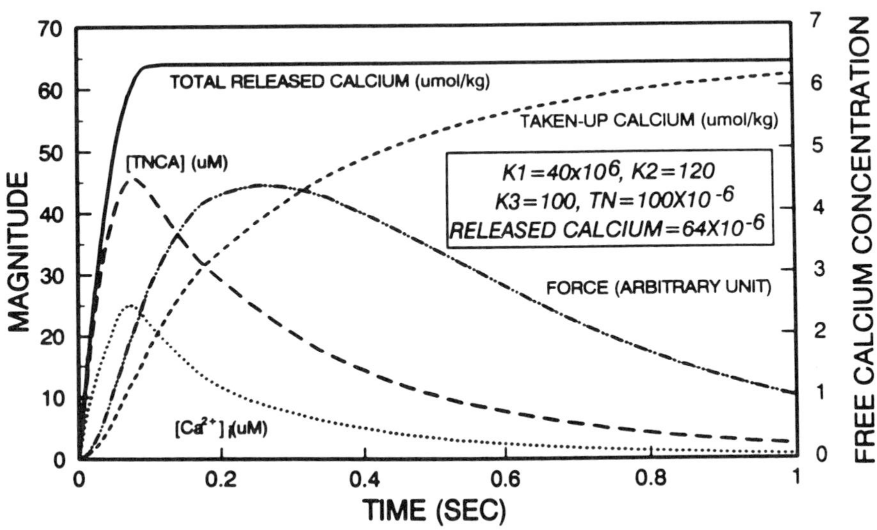

Fig. 3. Representative curves obtained in the simulation. Total released Ca (μmol/kg) is the integral of Ca^{++} release rate. Ca taken-up (μmol/kg) is the time integral of Ca^{++} uptake rate. [TnCa] (μM) is the concentration of Ca-bound troponin. $[Ca^{++}]_i$ (μM) is free Ca concentration (Ca transient). The left vertical coordinate shows the magnitudes of total Ca released and taken-up, and [TnCa]. The right vertical coordinates show the magnitude of $[Ca^{++}]_i$. The inset lists K1 (k_1), K2 (k_2), K3 (k_3), TN (total troponin) and released Ca.

Similar curves as shown in Figure 3 were obtained for various combinations of Ca and CB kinetic parameters. From these results, we found the following characteristics.

O_2 cost of Emax is hyperbolically related to Ka ($=k_1/k_2$), as shown in Figure 4. This cost increases sharply with decreases in Ka below the normal level, but decreases only moderately with increases in Ka above normal (eg, Ka = 2) (33). We found the underlying mechanisms to involve increases in the maximally activated twitch force despite no cooperativity (unity Hill coefficient) involved in Ca binding and CB attachment (33). If a Hill coefficient greater than unity is considered, a slightly greater sensitivity of O_2 cost of Emax to Ka would be expected [Namba, Takaki, Suga: unpublished].

The following simulation results are unpublished [manuscripts in preparation]. Briefly, O_2 cost of PVA increases sensitively with increases in g at a constant f, but decreases much less sensitively with increases in f at a constant g under a given combination of Ca transient and Ca kinetics. However, O_2 cost of PVA is not directly related to f and g when Ca transient and Ca kinetics vary.

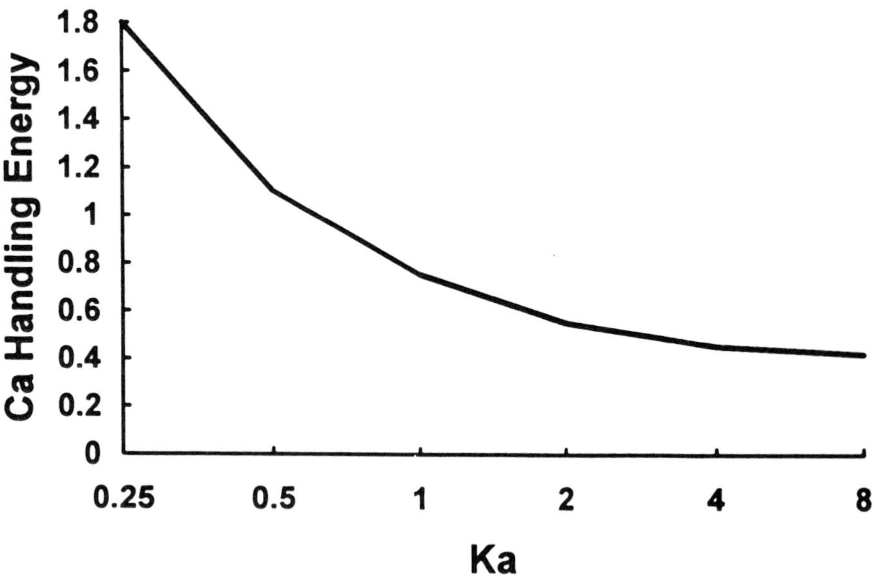

Fig. 4. A representative simulation result of Ca handling energy during excitation-contraction coupling for maintaining a constant contractile state. Ca handling energy decreases sharply with increases in Ka below its normal level of 2 but only slightly with further increases in Ka above normal.

O_2 cost of PVA increases with simultaneous increases in f and g under a given combination of Ca transient and Ca kinetics. However, O_2 cost of PVA decreases with changes in Ca kinetics despite constant f and g.

In contrast, O_2 cost of PVA is constant when Ca transient, Ca kinetics, f and g change proportionally, and relatively constant when only the magnitude of the Ca transient changes.

The reciprocal of the O_2 cost of PVA reflects the contractile efficiency of the contractile machinery (6,7,34). Therefore, the constancy and variability of the O_2 cost of PVA are inversely reflected in the contractile machinery. Thus, the constancy and variability of the contractile efficiency are, like those of the O_2 cost of PVA, closely related to the constancy and variability of Ca and CB kinetics.

In contrast to the contractile efficiency, economy of force time integral (FTI), defined as the ratio of FTI to energy consumption, often in terms of heat (35), changes inversely only with g and is completely independent of f and Ca kinetics as previously predicted by different theoretical approaches (36,37). The discrepancy of the behavior of the contractile efficiency and economy can be accounted for by the difference of the dimensions of these two terms. Contractile efficiency is dimensionless whereas economy has the dimensions of 1/-[speed]. If [speed] is normalized to [length], economy has the dimensions of [time]. This is the reason why these two mechanoenergetic efficacy indexes behave differently.

Implications

These simulation predictions may be related to the cellular mechanisms underlying the experimental observation that both O_2 costs of Emax and PVA are relatively constant in most cases of inotropic interventions but vary in some pathological states as discussed below.

The increased O_2 cost of Emax in stunned or acidotic hearts (22,23) may be explained partly by a decreased Ka and by futile Ca cycling. The significance of futile Ca cycling by the SR to increase PVA-independent Vo_2 despite decreased Emax has been beautifully demonstrated by ryanodine ($<$ 30 nM), which is known to make the Ca channel of the SR open-fixed and hence leaky to Ca^{++} (32). A decreased O_2 cost of PVA in stunned and acidotic hearts might be explained if CB kinetics were suppressed, but we have no evidence to support this hypothesis.

The virtually constant O_2 costs of PVA and Emax under most inotropic interventions such as catecholamines (16,26), digitalis (38), β blockers (19,20), Ca antagonists (39, our verapamil data (unpublished), pentobarbital sodium (21), fentanyl (40), non-ischemic transient Ca overload (41), capsaicin (42), new cardiotonic agents (18,27,28) and paired pulse stimulation (17) may be accounted for by simulation results of the case where all Ca and CB kinetics change proportionally or the case where only the peak Ca transient changes with all Ca and CB kinetics unchanged.

The constant O_2 cost of PVA over a wide range of heart rate may be related to proportional changes in Ca and CB kinetics as heart rate is varied.

The unchanged O_2 cost of PVA despite myocardial cooling and warming (29-41°C) may be related to proportional changes of Ca and CB kinetics under varied temperature. The nearly zero O_2 cost of Emax with cooling or warming may be accounted for if cooling increases Ka as a result of decreasing k_2 or warming decreases Ka as a result of increasing k_2 under the assumption of constancy of k_1, which is known to be dependent on Ca^{++} diffusion distance.

As for new cardiotonic agents including EMD 53998, which has a Ca sensitizing effect, O_2 cost of Emax was expected to decrease with the augmented responsiveness of the contractile machinery to Ca^{++}, but it did not (6,7,18,28,43). These agents were given to normal canine hearts, which probably had normal Ka. This might be one of the reasons that the O_2 cost of Emax was not significantly decreased with these Ca sensitizers. According to the simulation results, we expect a significant decrease in the O_2 cost of Emax if a Ca sensitizer is given to a failing heart with a decreased Ka (33). This expectation warrants a new experiment.

An increased O_2 cost of PVA in hyperthyroid rabbit hearts (24) could be accounted for by increased Ca and CB kinetics with a lesser effect on the duration of Ca transient. Both Ca and CB kinetics are known to be augmented as a consequence of phenotype changes of Ca pump and myosin ATPases in the hyperthyroid rabbit myocardium (35). The unchanged O_2 cost of PVA in hyperthyroid dog hearts despite evidence of increased CB cycling rate (25) may be related to the largely unchanged myosin isozyme and Ca-activated myosin ATPase activity. This discrepancy between the rabbit and the dog is probably accounted for by the constancy and variability of the O_2 cost of PVA as a function of Ca transient and Ca and CB kinetics, but not by the species difference.

CONCLUSIONS

Our experimental findings of the constancy and variability of the O_2 costs of PVA (mechanical energy) and Emax (contractility) of the left ventricle are summarized in relation to a variety of cardiac loading and inotropic conditions. Our simulation of myocardial calcium and crossbridge kinetics has provided possible mechanisms underlying these experimentally observed mechano-energetics under a variety of conditions.

ACKNOWLEDGEMENTS

Partly supported by Grants-in-Aid for Scientific Research (04454267, 04557041, 05221224, 05305007) from the Japan Ministry of Education, Science and Culture, and Research Grants for Cardiovascular Diseases (3A-2,4C-4) from the Japan Ministry of Health and Welfare.

REFERENCES

1. Suga, H. Theoretical analysis of a left ventricular pumping model based on the systolic time-varying pressure-volume ratio. IEEE Trans Biomed Eng BME-18:47-55, 1971.

2. Suga, H. Total mechanical energy of a ventricular model and cardiac oxygen consumption. Am J Physiol 236:H498-H505, 1979.

3. Suga, H., Sagawa, K., Shoukas, AA. Load independence of the instantaneous pressure-volume ratio of the canine left ventricle and effects of epinephrine and heart rate on the ratio. Circ Res 32:314-322, 1973.

4. Suga, H., Sagawa K. Instantaneous pressure-volume relationships and their ratio in the excised, supported canine left ventricle. Circ Res 35:117-126, 1974.

5. Suga, H., Hayashi, T., Shirahata, M. Ventricular systolic pressure volume area as predictor of cardiac oxygen consumption. Am J Physiol 240:H39-H44, 1981.

6. Suga, H. Ventricular energetics. Physiol Rev 70:247-277, 1990.

7. Suga, H., Goto, Y. Cardiac oxygen costs of contractility (Emax) and mechanical energy (PVA): New key concepts in cardiac energetics. In: Sasayama, S., Suga, H. (eds). Recent Progress in Failing Heart Syndrome. Springer-Verlag, Tokyo, pp 61-115, 1991.

8. Suga, H: Cardiac performance as viewed through the pressure-volume window. Circulation 88(suppl 4):I-C, 1993.

9. Suga, H., Goto, Y., Kawaguchi, O., Hata, K., Takasago, T., Saeki, A., Taylor, T.W. Ventricular perspective on efficiency. Bas Res Cardiol 88 (suppl 2):43-65, 1993.

10. Taylor, T.W., Goto, Y., Suga, H. Variable crossbridge cycling-ATP coupling accounts for cardiac mechanoenergetics. Am J Physiol 264:H994-H1004, 1993.

11. Goto, Y., Futaki, S., Kawaguchi, O., Hata, K., Takasago, T., Saeki, A., Nishioka, T., Taylor, T.W., Suga, H. Coupling between regional myocardial oxygen consumption and contraction under altered preload and afterload. J Am Coll Cardiol 21:1522-1531, 1993.

12. Hisano, R., Cooper, G. Correlation of force-length area with oxygen consumption in ferret papillary muscle. Circ Res 61:318-328, 1987.

13. Goto, Y., Slinker, B.K., LeWinter, M.M. Similar normalized Emax and O_2 consumption-pressure-volume area relation in rabbit and dog hearts. Am J Physiol 255:H366-H374, 1988.

14. Nozawa, T., Yasumura, Y., Futaki, S., Tanaka, N., Uenichi, M., Suga, H. The linear relation between oxygen consumption and pressure-volume area in left ventricle can be reconciled with the Fenn effect. Circ Res 65:1380-1389, 1989.

15. Taylor, T.W., Goto, Y., Suga, H. Myocardial mechanics and the Fenn effect determined from a cardiac muscle crossbridge model. Med Biol Eng Comput 31:377-383, 1993.

16. Suga, H., Hisano, R., Goto, Y., Yamada, O., Igarashi, Y. Effect of positive inotropic agents on the relation between oxygen consumption and systolic pressure-volume area in canine left ventricle. Circ Res 53:306-318, 1983.

17. Suga, H., Futaki, S., Tanaka, N., Yasumura, Y., Nozawa, T., Wu, D., Ohgoshi, Y., Yaku, H. Paired pulse pacing increases cardiac O_2 consumption for activation without changing efficiency of contractile machinery in canine left ventricle. Heart Vessels 4:79-87, 1988.

18. Futaki, S., Nozawa, T., Yasumura, Y., Tanaka, N., Suga, H. New cardiotonic agent, OPC-8212, elevates the myocardial oxygen consumption versus pressure-volume area (PVA) relation in a similar manner to catecholamines and calcium in canine hearts. Heart Vessels 4:153-161, 1988.

19. Suga, H., Goto, Y., Yasumura, Y., Nozawa, T., Futaki, S., Tanaka, N., Uenishi, M. O_2 consumption of dog heart under decreased coronary perfusion and propranolol. Am J Physiol 254:H292-H303, 1988.

20. Zhao, D.D., Namba, T., Araki, J., Ishioka, K., Takaki, M., Suga, H. Nipradilol depresses cardiac contractility and O_2 consumption without decreasing coronary resistance in dogs. Acta Med Okayama 47:29-33, 1993.

21. Namba, T., Takaki, M., Araki, J., Ishioka, K., Suga, H. Energetics of the negative and positive inotropism of pentobarbitone sodium in the canine left ventricle. Cardiovasc Res 28:557-565, 1994.

22. Ohgoshi, Y., Goto, Y., Futaki, S., Yaku, H., Kawaguchi, O., Suga, H. Increased oxygen cost of contractility in stunned myocardium of dog. Circ Res 69:975-988, 1991.

23. Hata, K., Goto, Y., Kawaguchi, O., Takasago, T., Saeki, A., Nishioka, T., Suga, H. Hypercapnic acidosis increases oxygen cost of contractility in the dog left ventricle. Am J Physiol 266:H730-H740, 1994.

24. Goto, Y., Slinker, B.K., LeWinter, M.M. Decreased contractile efficiency and increased nonmechanical energy cost in hyperthyroid rabbit heart: relation between O_2 consumption and systolic pressure-volume area or force-time integral. Circ Res 66:999-1011, 1990.

25. Suga, H., Tanaka, N., Ohgoshi, Y., Saeki, Y., Nakanishi, T., Futaki, S., Yaku, H., Goto, Y. Hyperthyroid dog left ventricle has the same oxygen consumption versus pressure-volume area (PVA) relation as euthyroid dog. Heart Vessels 6:71-83, 1991.

26. Ohgoshi, Y., Goto, Y., Kawaguchi, O., Yaku, H., Takaoka, H., Hata, K., Takasago, T., Suga, H. Epinephrine and calcium have similar oxygen cost of contractility. Heart Vessels 7:123-132, 1992.

27. Futaki, S., Goto, Y., Ohgoshi, Y., Yaku, H., Suga, H. Similar oxygen cost of myocardial contractility between DPI 201-106 and epinephrine despite different subcellular mechanisms of action in dog hearts. Heart Vessels 7: 8-17, 1992.

28. Hata, K., Goto, Y., Igarashi, Y., Yasamura, Y., Nozawa, T., Futaki, S., Takasago, T., Saeki, A., Taylor, T.W., Nishioka, T., Suga, H. Mechano-energetic effects of pimobendan in canine left ventricles. Comparison with dobutamine. Circulation 86:1291-1301, 1992.

29. Suga, H., Goto, Y., Igarashi, Y., Yasumura, Y., Nozawa, T., Futaki, S., Tanaka, N. Cardiac cooling increases Emax without affecting relation between O_2 consumption and systolic pressure-volume area in dog left ventricle. Circ Res 63:61-71, 1988.

30. Saeki, A., Goto, Y., Takasago, T., Nishioka, T., Suga, H. Effect of hyperthermia on left ventricular contractility and energetics in dog heart (abstract). Heart Vessels Suppl 8:66, 1992.

31. Nishioka, T., Goto, Y., Hata, K., Takasago, T., Saeki, A., Suga H. Mechanical vibration depresses cardiac contractility without affecting excitation-contraction coupling energy in isolated, blood-perfused dog heart (abstract). Heart Vessels Suppl 8:50, 1992.

32. Takasago, T., Goto, Y., Kawaguchi, O., Hata, K., Saeki, A., Nishioka, T., Suga, H. Ryanodine wastes oxygen consumption for Ca^{++} handling in the dog heart. A new pathological model. J Clin Invest 92:823-830, 1993.

33. Namba, T., Takaki, M., Araki, J., Ishioka, K., Akashi, T., Zhao, L.Y., Matsushita, T., Ito, H., Fujii, w., Matsubara, H., Suga, H. Ca^{++} sensitivity of contractile machinery and Ca^{++} handling energy. Simulation. Jpn Heart J 34:601-616, 1993.

34. Takaki, M., Namba, T., Araki, J., Ishioka, K., Ito, H., Zhao, L.Y., Zhao, D.D., Liu M., Fujii, W., Suga, H. How to measure cardiac energy expenditure. In: Piper, H.M., Preusse C.J. (eds). Ischemia-reperfusion in cardiac surgery. Kluwer Academic Pub, The Netherlands 403-419, 1993.

35. Alpert, N.R., Mulieri, L.A. Determinants of energy utilization in the activated myocardium. Fed Proc 45:2597-2600, 1986.

36. Suga, H., Taylor, T.W. Myocardial efficiency and economy in Huxley's 1957 crossbridge model. Jpn Heart J 32:827-834, 1991.

37. Suga, H., Goto, Y., Hata, K., Takasago, T., Saeki, A., Taylor, T.W., Nishioka, T. Constant efficiency versus variable economy of cardiac contraction. Jpn Heart J 33:213-227, 1992.

38. Wu, D., Yasumura, Y., Nozawa, T., Tanaka, N., Futaki, S., Ohgoshi, Y., Yaku, H., Suga, H. Effect of ouabain on the relation between left ventricular oxygen consumption and systolic pressure-volume area (PVA) in dog heart. Heart Vessels 5:17-24, 1989.

39. Burkhoff, D., Yue, D.T., Oikawa, R.Y., Franz, M.R., Schaefer, J., Sagawa, K. Influence of ventricular contractility on non-work-related myocardial oxygen consumption. Heart Vessels 3:66-72, 1987.

40. Ishioka, K., Takaki, M., Namba, T., Araki, J., Suga, H. Fentanyl depresses myocardial contractility and oxygen consumption in blood-perfused isolated canine left ventricle (abstract). Japan-USA Heart Energy Workshop Program and Abstracts, Okayama, 1994.

41. Araki, J., Takaki, M., Namba, T., Mori, M., Suga, H. Transient Ca overload suppresses contractility and excitation-contraction coupling energy in excised cross-circulated canine left ventricle. Japan-USA Heart Energy Workshop Program and Abstracts, Okayama, 1994.

42. Takaki, M., Akashi, T., Ishioka, K., Kikuta, A., Mataubara, H., Yasuhara, S., Fujii, W., Suga, H. Effects of capsaicin on mechanoenergetics of excised cross-circulated canine left ventricle and coronary artery. J Mol Cell Cardiol, in press, 1994.

43. deTombe, P.P., Burkhoff, D., Hunter, W.C. Effects of calcium and EMD 53998 on oxygen consumption in isolated canine hearts. Circulation 86: 1945-1954, 1992.

2. Kinetic Property of Cardiac Myosin In Vitro

Seiryo Sugiura, Hiroshi Yamashita, Masataka Sata, Hideo Fujita, Shin-ichi Momomura, Takashi Serizawa, Haruo Sugi

INTRODUCTION

Cardiac mechanics has been studied at various levels of the circulatory system. In systems physiology, the heart is considered a pump whose property is characterized by an input-output relation. However, recent advances in biochemistry and molecular biology have revealed aspects of the internal structure of this black box at the molecular level (1). Often, however, the functional significance of these microscopic findings is not clear. In other words, there is a gap between microscopic and macroscopic findings.

A powerful approach to this problem is the recently developed *in vitro* motility assay (2,3,4,5,6,7). With this technique we can study the mechanical property of contractile proteins under controlled conditions and thus establish a direct relation between mechanical function and structure or biochemistry. In this chapter, we will briefly review *in vitro* motility assay techniques developed by various researchers and present our findings for cardisac myosin (8,9,19).

METHODS

Preparation of Proteins

Cardiac myosin was obtained from left ventricular muscle of rat, rabbit or hamster according to the method of Katz et. al. (11) with some modifications. All purification procedures were carried out at 4°C. The myosin solution was stored in an ice-cooled container and used for the determination of ATPase activity (by the method by Yazaki and Raben (12)) and the *in vitro* motility assay within 48 hour after purification. Myosin isoform analysis was performed by pyrophosphate polyacrylamide gel electrophoresis according to Martin et al (13). After being fixed and stained, the gel was scanned by a laser densitometer and the area under each peak was measured to determine the isoform distribution. The percentage of α-myosin heavy chain (%α-MHC) was determined by the following formula: $\%\alpha\text{-MHC} = \%V_1 + \%V_2/2$.

Monomeric (G-) actin was purified from an acetone powder by the method of Spudich and Watt. G-actin was incubated in a high ionic strength buffer with rhodamine-phalloidin to form fluorescently labelled actin filament (F-actin).

Animal Models

To study the relationship between cardiac myosin isoforms (V_1,V_3) and its mechanical property, the following animal models were used: 1)Rabbit hyperthyroidism model; to achieve myosin isoform redistribution from V_3 to V_1, hypothyroidism was induced in the adult rabbit. 2)Rat hypothyroidism model; to obtain V_3 myosin isoform, hypothyroidism was induced in 12-week old male Wistar rats by adding methimazole to their drinking water. V_1 myosin isoform was obtained from 3-week old rat hearts. 3)Cardiomyopathic hamster model; the relative amount of V_3 isoform in cardiac myosin obtained from cardiomyopathic hamster (Bio 14.1 strain) is greater than that from control (F1B strain). Accordingly, we compared the myosin from these animals.

In Vitro Motility Assay

The essence of the *in vitro* motility assay is to reconstitute actomyosin sliding with myosin (and actin) purified from muscle tissue. Active interaction of actomyosin takes place only in a physiological buffer solution which excludes the use of an electron microscope. Therefore, a technique to visualize actomyosin sliding under light microscopy is the key in this assay. For this purpose, two types of *in vitro* motility assay systems are currently used widely.

Algal Actin Based Motility

In this type of assay, actin is obtained from fresh water alga *Nitellopsis obtusa* (2,3). This alga is made up of internodal cells connected in series and each cell, usually several centimeters long, has well organized rows of actin cables on its inner surface. These actin cables interact with algal myosin attached to organelles and generate motive force for cytoplasmic streaming. By cutting open both ends of the cell and perfusing the inside, actin cables are exposed. Myosin coated beads suspended in Mg-ATP solution (2mM, ATP, 5mM EGTA, 6mM $MgCl_2$, 70mM KOH, 30mM PIPES, and 200mM sorbitol pH 7.0) were perfused into actin donor cells prepared in the above manner. The sliding movement of the bead dragged by the actomyosin interaction was observed under a light microscope. The observed movement was always linear reflecting the well organized structure of the algal actin cables.

Fluorescent Actin Sliding Assay

In this type of assay (4), both actin and myosin are obtained from muscle tissue. First, myosin was fixed on a glass coverslip. Onto this myosin layer fluorescently labelled actin filaments were introduced with a buffer containing MgATP. The sliding movement of actin filament on the myosin layer was observed with an inverted fluorescence microscope. In this assay, the actin filament followed a winding path because the arrangement of myosin was random.

Force Movement Assay

Under physiological condition, cardiac muscle contracts against a load. To obtain information on the force generating property of contractile proteins, different types of *in vitro* assay systems have been introduced. These techniques are based on either fluorescent actin motility (5) or algal actin based motility (6,7). Of these, we used a force-movement assay technique developed by Suga et. al. (7). This technique based on the algal actin assay uses a centrifuge microscope. The microscope is equipped with a revolving stage and strobe light which illuminates the specimen in synchrony with the rotation of the stage. The algal cell preparation described above is placed in a cuvette and then mounted on the stage. By rotating the stage we can apply a centrifugal force which serves as a load to the bead movement. The amount of force is calculated by the following equation:

$$\mathbf{P} = \Delta\sigma\cdot\mathbf{V}\cdot\mathbf{r}\cdot\omega^2$$

where $\Delta\sigma$ is the difference in density between the bead and the surrounding medium, V is the bead volume, r is the effective radius of centrifugation, and ω is the angular velocity of the rotor. Further details of the apparatus are discussed in ref. 7. The advantage of this technique is that we can obtain a steady state force-velocity relation *in vitro*. In addition, by reversing the direction of force, we can apply a negative load to the sliding bead.

Experiments were performed at room temperature. An image of the bead moving under constant load was recorded by a video camera and a video cassette recorder system. The bead movement was analyzed on replay of the video tape and the velocity was measured.

RESULTS

Velocity of Actomyosin Sliding Under Different Conditions

In the following part of this review, the data were obtained primarily from the algal actin based motility assay. Under control condition (pH 7.0, ATP 2mM), myosin coated beads moved smoothly along the actin cables at a constant velocity. The average velocity of the bead coated with euthyroid rabbit myosin was 0.31±0.11 μm/s and that with 3-week old rat myosin (mainly V_1) was 1.0±0.26 μm/s (8).

In Figure 1, the sliding velocity of the bead coated with euthyroid rabbit cardiac myosin is plotted as a function of ATP concentration of the medium (8). The movement is dependent on [ATP]. However, above a certain level of ATP, the velocity remains constant.

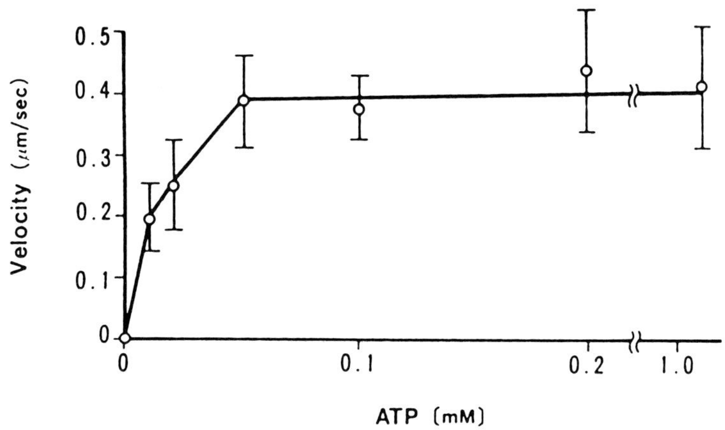

Fig. 1. Sliding velocity of a bead coated with rabbit cardiac myosin as a function of ATP concentration of the medium. Reprinted with permission from ref. 8.

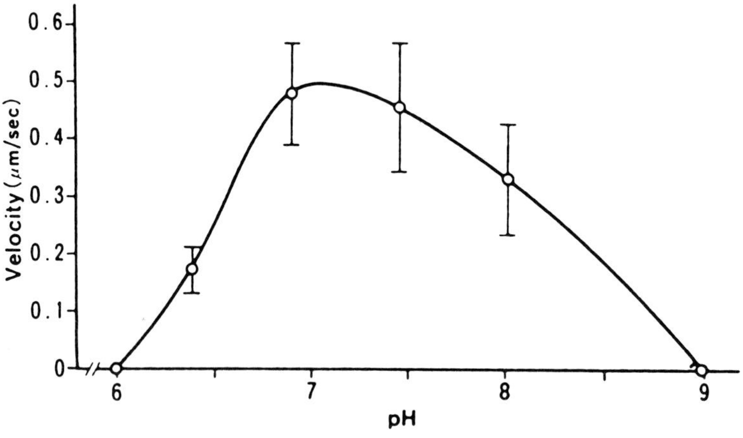

Fig. 2. The effect of pH on the sliding velocity of a bead coated with rabbit cardiac myosin. Reprinted with permission from ref. 8.

The relation between the sliding velocity and the pH of the medium is plotted for euthyroid rabbit myosin in Figure 3 (8). The velocity has a parabolic relation to pH with a maximum around pH 7.0. The decline in sliding velocity was steeper on the acidic side suggesting a strong cardiac depressant effect of acidosis.

These data are similar to what has been observed with muscle preparations indicating that the reconstituted actomyosin sliding in this assay system is physiological in nature.

Myosin Isoform and Sliding Velocity

Using the animal models described above, we could obtain cardiac myosin having various isoform distributions. We studied both biochemical and mechanical properties of these myosin samples. Figure 3 depicts the relation between myosin isoform distribution (% α-MHC) and Ca^{++}-ATPase activity. The data shown are for rabbit and hamster. There was a good correlation between these parameters.

The relation between mechanical property and molecular structure is shown in Figure 4 in which the sliding velocity is plotted as a function of % α-MHC (10). We also found a good correlation between these parameters. These results suggest that under control condition the sliding velocity of actomyosin is determined by the molecular structure of myosin. There are, however, two things to be noted: 1) because the

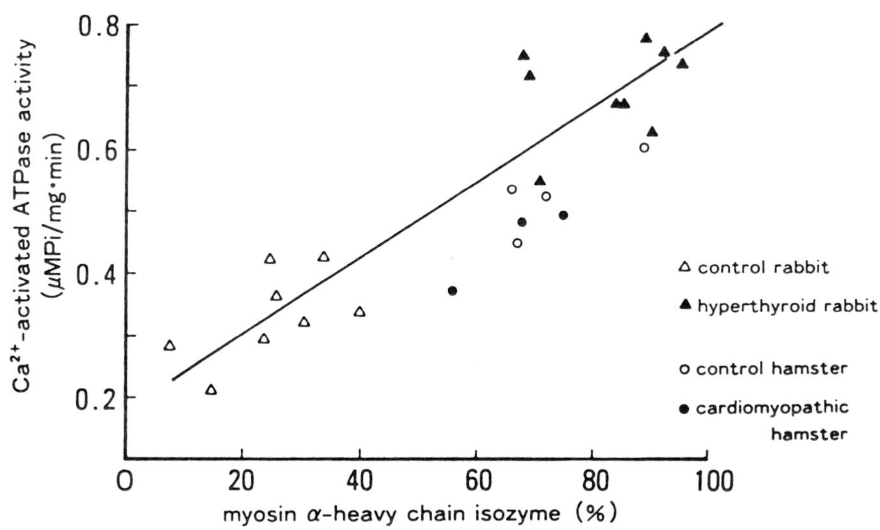

Fig. 3. Relationship between Ca^{++}-activated ATPase activity and myosin isoform composition.

range of % ∝-MHC for hamster myosin was limited, we cannot be sure whether the relation between % ∝-MHC and velocity falls on the same relation for rabbit myosin; 2) Although there was a strong positive correlation, the relation is not necessarily linear. In fact we have recently shown a non-linear relation for a mixture of V_1 and V_3 myosin obtained from rat cardiac muscle (14).

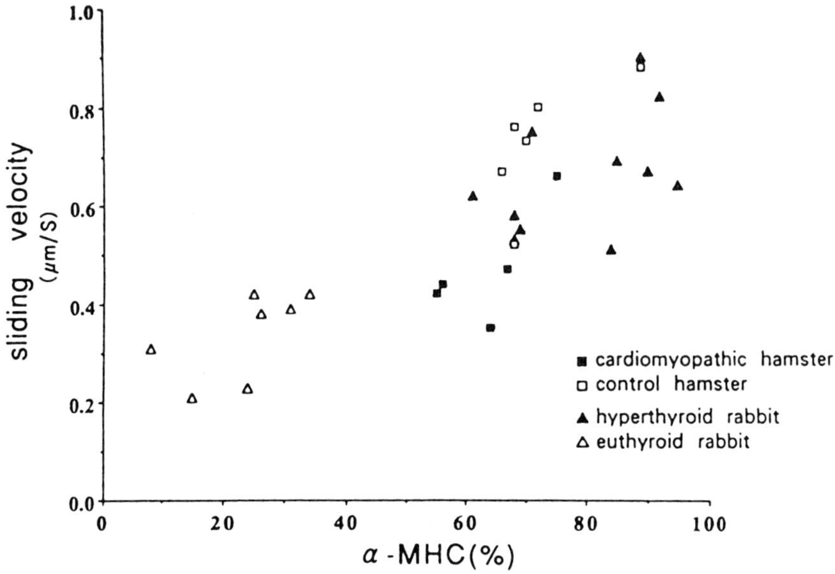

Fig. 4. Relationship between sliding velocity and myosin isoform composition. Reprinted with permission from ref. 10.

Steady-State Force-Velocity Relation of the Actomyosin Sliding *In Vitro*

Under constant load, the bead moved at a constant velocity for many seconds indicating that a steady-state was established. In Figure 5, the steady-state force-velocity relation is shown for hyperthyroid rabbit cardiac myosin. As the load was increased, the velocity decreased in a hyperbolic manner in the low load range. However, in the high load range the shape of the relation deviated from the hyperbola. This characteristic shape is similar to what has been reported for skeletal muscle myosin (7). The significance of this finding is two fold. First, actomyosin sliding reconstituted *in vitro* can generate force. Second, the reconstituted system has a physiological force-velocity relationship.

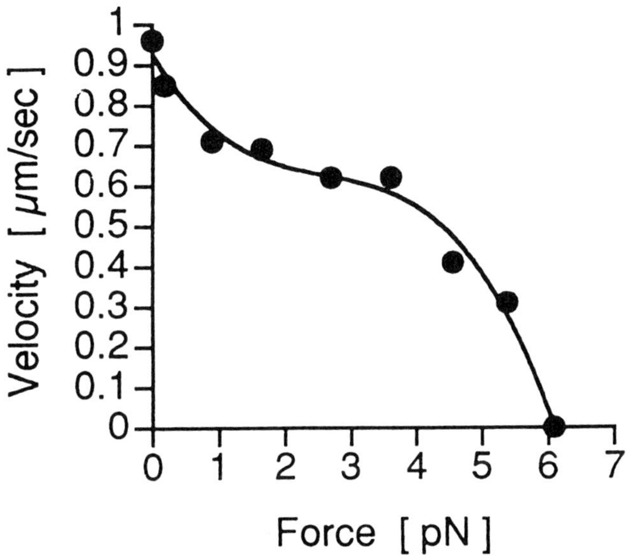

Fig. 5. Steady-state force-velocity relation of a bead coated with hyperthyroid rabbit cardiac myosin.

DISCUSSION

In Vitro Motility Assay

To date, various types of *in vitro* motility assay techniques have been introduced. With these techniques, we can study mechanical properties of the actomyosin interaction in a simple system under controlled conditions. As shown in this article, reconstituted actomyosin sliding retains its physiological nature. The currently used assay techniques are classified into category. The fluorescent actin sliding technique can be utilized to study both myosin and actin of desired species. On the other hand, the algal actin based assay enables us to obtain steady-state conditions lasting several seconds.

Vmax and Myosin Isoform

Barany showed that unloaded shortening velocity of muscle (V_{max}) correlates with the ATPase activity of myosin obtained from that muscle (15). Since then similar results have been reported for various types of muscle including cardiac (16). Because the force required to drag a bead in the medium is much smaller than that generated by a single crossbridge (about 1 pN), sliding velocity observed *in vitro* can be considered as an equivalent of V_{max} (2). In fact, sliding velocities reported so far for various muscle myosin have correlated well with ATPase activity (9,17).

Force-Velocity Relation *In Vitro*

It has been shown that isolated actin and myosin can generate active force in several types of *in vitro* force-movement assay systems. Yanagida et. al. using a compliant glass microneedle as a force transducer, measured the force generated by an actin filament sliding over a myosin coated glass surface (5). Force-velocity relations have also been reported in various other assay techniques (6,7). However, the only technique with which we can measure a steady-state force-velocity relation comparable to those obtained in muscle preparations is the centrifuge microscope system (7).

Although the observed force-velocity relation resembled that of intact muscle, it differed in some respects. In our experiments with rabbit cardiac myosin, the maximum force was about 5 to 10 pN. If we assume that each crossbridge generates about 1pN, the number of crossbridges participating in the bead movement is on the same order. Taking this fact into account, the force-velocity relation *in vitro* may reflect the kinetic property of each crossbridge. The shape of the *in vitro* force-velocity relation differed from that of intact muscle especially in the high load region. To account for this difference, we propose a conceptual model.

During muscle contraction, the myosin head is believed to repeat attachment and detachment cycles with actin (Fig. 6A). In intact muscle, in which large numbers of myosin heads work together, the discrete nature of each interaction will be smoothed out. However, if the number of myosin heads is limited, as it is in the *in vitro* situation, the sliding movement will become discontinuous (repetitions of advance and pause) and the average velocity will be reduced. Furthermore, if centrifugal force is applied opposite to the direction of movement, the bead will be pushed backward during the detachment period resulting in a further reduction in average velocity (Fig. 6B). The reduction will be larger under high load and thus the deviation of force-velocity relation from the hyperbola will be accentuated in the high load range. We believe that the characteristic shape of force-velocity relation *in vitro* originates mainly from the small number of crossbridges participating in sliding and that the *in vitro* force-velocity relation reflects a physiologic property of actomyosin interaction.

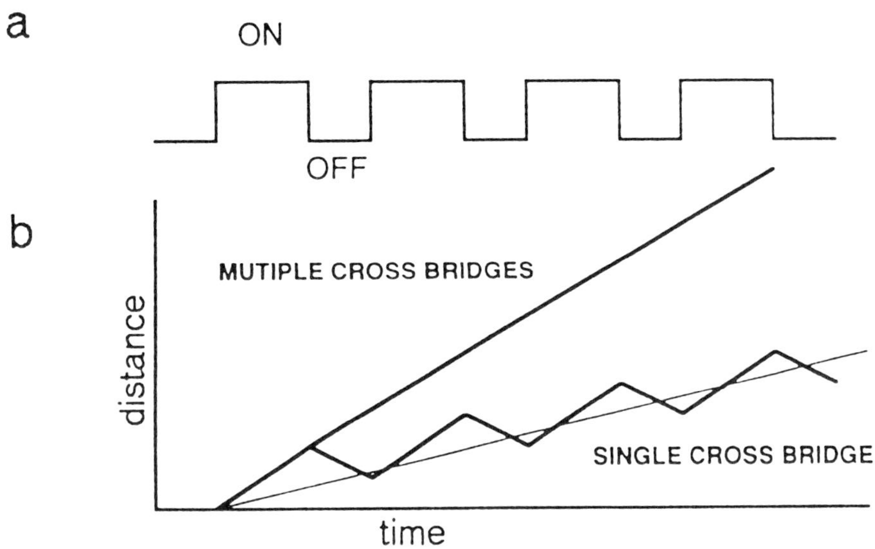

Fig. 6. A: Diagram of crossbridge cycle. Each crossbridge repeats attachment and detachment cycle. B: Sliding movement of bead driven by single and multiple crossbridges.

In summary, we have utilized *in vitro* motility assay techniques to study the mechanical property of cardiac myosin under various conditions for different myosin isoforms. Although these findings were anticipated based on previous experiments with muscle preparations, this is the first presentation of such direct evidence at the molecular level.

REFERENCES

1. Lompre, A.M., Nadal-Ginard, B., Mahdavi, V. Expression of the cardiac ventricular α- and β-myosin heavy chain genes is developmentally and hormonally regulated. J Biol Chem 259:6437-6446, 1984.

2. Sheetz, M.P., Spudich, J.A. Movement of myosin-coated fluorescent beads on actin cables *in vitro*. Nature 303:31-35, 1983.

3. Shimmen, T., Yano, M. Active sliding movement of latex beads coated with skeletal muscle myosin on Chara actin bundles. Protoplasma 121:132-137, 1984.

4. Kron, S.J., Spudich, J.A. Fluorescent actin filaments move on myosin fixed to a glass surface. Proc Natl Acad Sci USA 83:6272-6276, 1986.

5. Kishino, A., Yanagida, T. Force measurements by micromanipulation of a single actin filament by glass needles. Nature 334:74-76, 1988.

6. Chaen, S., Oiwa, K., Shimmen, T., Iwamoto, H., Sugi, H. Simultaneous recordings of force and sliding movement between a myosin-coated glass microneedle and actin cables *in vitro*. Proc Natl Acad Sci USA 334:74-76, 1989.

7. Oiwa, K., Chaen, S., Kamitsubo, E., Shimmen, T., Sugi, H. Steady-state force-velocity relation in the ATP-dependent sliding movement of myosin-coated beads on actin cables *in vitro* studied with a centrifuge microscope. Proc Natl Acad Sci USA 87:7893-7897, 1990.

8. Sugiura, S., Yamashita, H., Serizawa, T., et. al. Active movement of cardiac myosin on Characeae actin cables. Pflugers Arch 421:32-36, 1992.

9. Yamashita, H., Sugiura S., Serizawa, T. et. al. Sliding velocity of isolated rabbit cardiac myosin correlates with isozyme distribution. Am J Physiol 263:H464-H472, 1992.

10. Yamashita, H., Sugiura, S., Sata, M. et. al. Depressed sliding velocity of isolated rabbit cardiac myosin from cardiomyopathic hamsters. Evidence for an alteration in mechanical interaction of actomyosin. Mol Cell Biochem 119:79-88, 1993.

11. Katz, A.M., Repke, D.I., Rubin, B.B. Adenosinetriphosphatase activity of cardiac myosin: comparison of the enzymatic activities and activation by actin of dog cardiac, rabbit cardiac, rabbit white skeletal and rabbit red skeletal muscle myosins. Circ Res 19:611-621, 1966.

12. Yazaki, Y., Raben, M.S. Cardiac myosin adenosinetriphosphatase of rat and mouse-distinct enzymatic properties compared with rabbit and dog cardiac myosin. Circ Res 35:15-23, 1974.

13. Martin, A.F., Pagani, E.D., Solaro, R.J. Thyroxine induced redistribution of isozyme of rabbit ventricular myosin. Circ Res 50:117-124, 1982.

14. Sata, M., Sugiura, S., Yamashita, H., et. al. Dynamic interaction between cardiac myosin isoforms modifies velocity of actomyosin sliding *in vitro*. Circ Res 73:696-704, 1993.

15. Bárány, M. ATPase activity of myosin correlated with speed of muscle shortening. J Gen Physiol 50:197-218, 1967.

16. Cappelli, V., Bottinelli, R., Poggesi, C. et. al. Shortening velocity and myosin and myofibrillar ATPase activity related to myosin isoenzyme composition during postnasal development in rat myocardium. Circ Res 65:446-457, 1989.

17. Sheetz, M.P., Chasan, R., Spudich, J.A. ATP-dependent movement of myosin *in vitro*: characterization of a quantitative assay. J Cell Biol 99:1867-1871, 1984.

3. Thin Filament Regulation and Cardiac Energetics

R. John Solaro

INTRODUCTION

It is now apparent that variation in the response of myofilaments to Ca^{++} represents one way cardiac muscle cells may regulate the amount of energy required to do pressure-volume work (1,2). Using a plausible model of excitation contraction coupling and mechanoenergetics, Namba et al. (2) quantified the extent to which changes in the Ca^{++}-affinity of troponin C (TnC), the Ca^{++}-receptor activating myofilament activity, may affect the total amount of Ca^{++} release required to develop contractile force. An important conclusion of this analysis is that with an increase in the affinity of TnC for Ca^{++}, the same contractile force is achievable with a lower amount of released Ca^{++}. This reduces the amount of Ca^{++} that cycles and thus the amount of energy expended for Ca^{++} transport. This is a straight-forward mechanism for a possible impact on energetics of altered myofilament response to Ca^{++}. However, changes in the affinity of TnC for Ca^{++} represent one of several possibilities for altering the response of myofilaments to Ca^{++}. It is now clear that control of the actin-myosin reaction is not only through Ca^{++}-binding to TnC, but also through cooperative/allosteric processes involving all of the main myofilament proteins:actin, myosin, tropomyosin (Tm), troponin T (TnT), troponin I (TnI) and TnC. In this chapter, I will summarize these mechanisms and give some examples of how they might vary physiologically and pathophysiologically. It will be seen that the control process is complex, involves a number of protein-protein interactions and is subject to modulation by non-covalent and covalent modifications as well as the mechanical state of the myofilaments.

CHEMOMECHANICAL COUPLING, AND SWITCHING ON OF THIN FILAMENTS BY CA^{++}.

It is useful to begin to analyze control of the actin-myosin reaction by considering current concepts of the relation between hydrolysis of ATP by myosin and actomyosin and its relation to force and shortening. Crossbridges appear to be in one of the following three states with regard to their interaction with actin (3): 1) a blocked state, in which binding to the thin filament is blocked by troponin (Tn)-Tm; 2) a weak binding state, in which crossbridges react with the thin filament in a rapid "on-off steady state" but do not generate force; and, 3) a strong binding state that generates force and, as will be discussed, cooperatively turns on the thin filament. In the absence of Tn-Tm or at least in the absence of TnI-TnC, there is an unbridled reaction cycle between crossbridges and actin that results in ATP hydrolysis and contraction. During the hydrolysis of ATP, the crossbridge goes through several phases in a reaction sequence in which each step is believed to be associated with a mechanical state of the myofilaments. Intermediates in the reaction sequence and the transition between weakly bound states and force generating states are depicted in Figure 1 according to the scheme of Millar and Homsher (4). Force and shortening are proportional to the fraction of crossbridges in the strong binding state. In turn, the fraction of strong crossbridges is determined by the extent to which the thin filament is reactive or "turned on" as it is commonly described. In the complete absence of Tn-Tm it is apparent that the thin filament is fully reactive. When Tn and Tm are present on the thin filament, the transition from the weakly bound state to a strong binding state is triggered by Ca^{++}. The step in the reaction sequence most likely controlled by Ca^{++}, according to the Millar and Homsher (4) model depicted in Fig. 1, is the slow step involving isomerization of a weak binding state, actin-myosin-ADP-Pi. Results from studies in cardiac muscle using sinusoidal analysis also agree that this step is the force generating step in heart myofilaments (5). Accordingly during diastole, in which Ca^{++} is limiting, weakly bound or blocked crossbridges accumulate and myofilaments are in a resting condition.

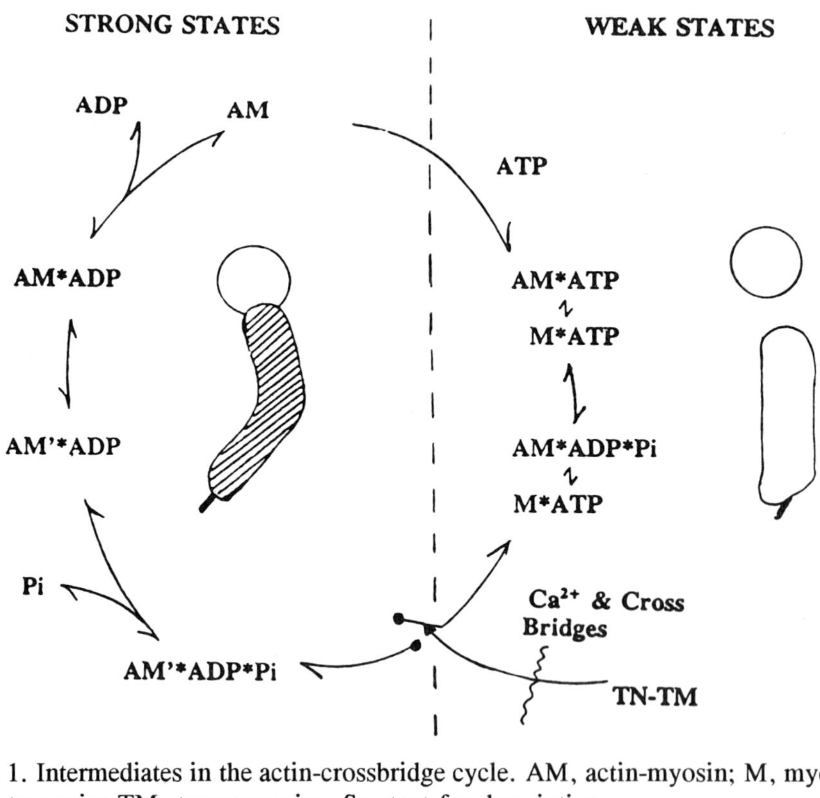

Fig. 1. Intermediates in the actin-crossbridge cycle. AM, actin-myosin; M, myosin; TN, troponin; TM, tropomyosin. See text for description.

Switching the Thin Filament "On" Through Ca-TnC

There is ample evidence that the reaction of Ca^{++} with TnC triggers activation of force generating states of the myofilaments (6). A current unresolved issue is just what is regulated when Ca^{++} binds to TnC. There are several main possibilities and there is evidence for each (5,7-9). The first is that blocked crossbridges are able to react with actin upon a Ca^{++}-dependent removal of a steric hindrance or an alteration in actin activity. These crossbridges would contribute to the total population of crossbridges able to undergo the weak to strong transition and presumably enter the cycle as weakly bound crossbridges. The idea that Tm movement is a key element in switching on the thin filament is a relatively old idea now (10), but has received support from more recent studies (11,12). It remains to be seen whether the movement analyzed by 3-D reconstructions involves only Tm or both Tm and Tn (13). An important consideration is whether the Tm or Tm-Tn movement removes a steric block or works by altering the activity of actin. Tm, when added to actin, may either inhibit or stimulate actomyosin ATPase activity (14). Moreover, Bremel and Weber (15) showed that Tm stimulates actin activated myosin ATPase activity in the absence of Ca^{++} depending on the number of strong crossbridges, which was varied by varying the MgATP concentration. These actions of Tm are difficult to explain by a steric effect and are more readily explained by allosteric effects on actin structure. Thus, it seems likely that a model restricted to the idea that movement of Tm or Tn permits binding of myosin heads is probably too simple and some allosteric effects on actin structure are also possible. Which effect predominates in the myofilament lattice at body temperature is not clear.

Possible steric and allosteric actions of Tm are also couched in evidence relating to opposing ideas on the way Ca^{++} switches on crossbridge activity. One possibility is that Ca^{++} acts as a switch and promotes the transition from the weak to the strong state in an "all or none" fashion. In this case Tm acts to enhance the affinity of crossbridges for actin, and relative activation of the myofilaments by Ca^{++}, say at the 50% level would involve 50% of the crossbridges "recruited" to the strong state. An opposite view is that Ca^{++} increases the rate of transition from the weak to the strong state in a "graded" fashion (16). In this case, 50% relative activation of the myofilaments could involve all of the crossbridges in an activated state, but with the forward rate of transition between weak and strong states at some sub-maximal level (16). A choice between these possibilities is difficult in that the conclusions from particular experiments are dependent on the methods used and may be model dependent. One method used to distinguish between these possibilities is to subject skinned fibers to a quick release, break all crossbridges, and follow with a quick restretch. In such experiments tension recovers with a measurable time course, and the rate of tension recovery (Ktr) is Ca^{++} dependent, especially at levels of Ca^{++} above that for half-maximal activation of steady-state force (17). An effect of Ca^{++}

on the rate of transition between weak and strong crossbridge states would predict that Pi release would be Ca^{++} dependent. Yet measurement of the kinetics of this step with the aid of caged Pi and measurement of kinetics of force transients associated with Pi release (K_{Pi}) have been interpreted as a demonstration of either no effect of Ca^{++} on the Pi release step (5) or a relatively small effect (17) that cannot account for the effects of Pi on steady-state pCa-force relations. On the other hand sinusoidal analysis shows the rate constants are not Ca^{++} dependent (18), suggesting crossbridge recruitment. The truth about the "graded" or "all or none" hypotheses has a bearing on the perception of the actions of Tm in as much as in one case ("all or none"), Tm could sterically or allosterically affect a pre-force generating state of the crossbridge in a reaction which is strongly Ca^{++}-dependent. In the case of a "graded" mechanism, the idea is that the crossbridges need not be in this blocked state, but in a weak state, whose affinity for actin, is not greatly affected by Ca^{++}. In the "graded" hypothesis, regulation by Ca^{++} occurs mainly by increasing the rate of transition from weak to strong binding states.

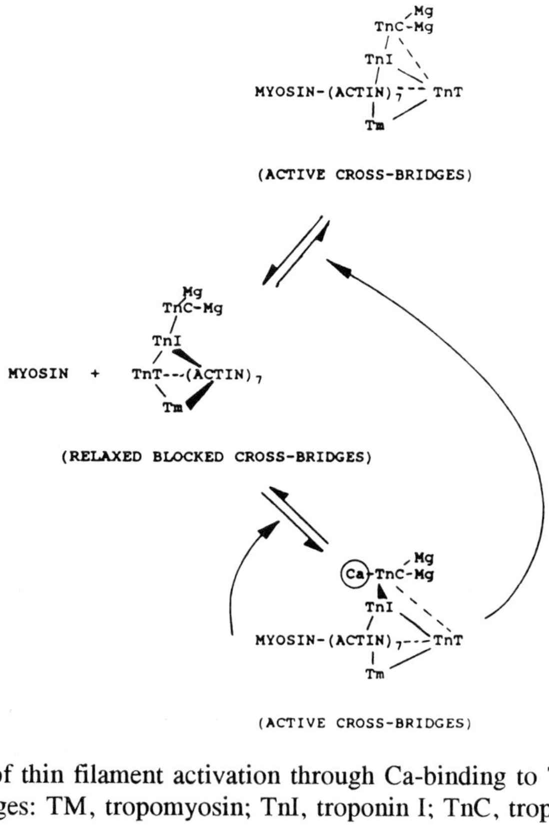

Fig. 2. Modes of thin filament activation through Ca-binding to TnC and through strong crossbridges: TM, tropomyosin; TnI, troponin I; TnC, troponin C. See text for discussion.

Whether "all or none" or "graded" in terms of activation of the thin filament, there is an emerging picture of how the reaction of Ca^{++} with TnC is transduced to affect the reactivity of actin. Figure 2 illustrates the molecular processes that are believed to occur in triggering the actin-myosin crossbridge reaction and shows the thin filament in "on" and "off" states. The "off" state is associated with low cytosolic Ca^{++}, there are weak interactions between TnC and TnI and possibly between TnC and TnT, and TnI binds to actin strongly (19). The "off" state may also be characterized by a poorly understood interaction of TnI with TnT, and a potential interaction of TnT with actin (20). These interactions, which are transmitted to Tm through TnT, hold Tm in a conformational state or location that alters actin activity such that the actin-crossbridge reaction is impeded. The strong interaction of TnI with actin in relaxing conditions also is likely to inhibit the reaction of crossbridges and has been speculated to produce the "blocked" state (3,7,9). As depicted in Figure 2, these inhibitory reactions are reversed when Ca^{++} binds to TnC and result in an activation of myofilament activity. A key reaction central to the release from inhibition is the tight binding of TnC with TnI upon Ca^{++}-binding to the regulatory site of TnC. The topology of the interactions are presently being sorted out. An interesting feature of the cardiac variant of TnI (cTnI) is an amino terminal extension of some 32 amino acids. This stretch of amino acids contains serines that are substrates for protein kinase A (21). We have shown using NMR that the interaction of TnI with TnC is anti-parallel, that is the N domain of cTnC is opposite the C domain of cTnI (22). Moreover, we have also shown that removal of the N domain of cTnI has no effect of Ca^{++} activation of myofibrillar ATPase activity or its inhibition by acidic pH (23). Our hypothesis is that the unique N domain of cTnI is important only when it is phosphorylated. We shall see later that phosphorylation of amino acids in the unique N domain (Ser 23/24) and just downstream (Ser 43/45) affect the Ca^{++}-sensitivity and maximum actomyosin ATPase rate. This latter finding indicates that the state of TnI may affect the ability of the thin filament to turn on even in saturating levels of Ca^{++}. Below, I describe evidence that, in fact, Ca-TnC alone cannot fully turn on the thin filament.

Switching the Thin Filament "On" Through Strong Crossbridges

Apart from this mechanism by which the actin-myosin crossbridge reaction is triggered by Ca^{++} binding to Tn, there is now substantial evidence that Ca^{++} may, in fact, act as a co-factor for a primary mechanism to turn on the thin filament, which involves binding of the crossbridge itself to the thin filament.

It is now clear that in striated muscle, myosin is not only a molecular motor but is also involved in activation of the thin filament. This action of myosin heads was first clearly demonstrated in vitro in the studies of Bremel and Weber (15), who showed that rigor crossbridges could, in the presence of Tm or Tn-Tm, cooperatively turn on the thin filament, even in the absence of Ca^{++}, as determined from measurement of acto-S-1 or acto-HMM ATPase activity. Later experiments by Williams et al (24) showed that Ca^{++} could not fully turn on the thin filament whereas addition of NEM-modified S-1 (an analogue of the strong crossbridge that does not split ATP) could. Binding of S-1 was subsequently shown to be cooperative by Green and Eisenberg (25) even in the presence of Ca^{++}. On the other hand in the case of cardiac thin filaments and skinned fiber preparations, the binding of Ca^{++} to TnC is weakly cooperative or non-cooperative (6). That binding of S-1 affects further binding of S-1 was also shown by the experiments of Ishii and Lehrer (26), who probed the number of active Tm units using a fluorescent tag reporting the "on" configuration. These experiments showed that turning on of the thin filament is largely dependent on S-1 binding and it is relatively weakly affected by Ca^{++} binding to Tn. The cooperative turning on of the thin filament lies in two mechanisms. One is the fact that Tm covers 7 actins so that with binding of S-1, 7 actins are activated. The other is that the cooperative binding of S-1 requires end-to-end interactions between adjacent Tm molecules. Thus, near neighbor interactions involving functional units consisting of actin:Tm:Tn in a 7:1:1 ratio are also involved. The size of the functional unit may however extend over a longer stretch of the thin filament as demonstrated by experiments of Metzger and Moss (27), who showed that extraction of about one-third of the Tn molecules from the thin filament could fully activate skinned fibers. Moreover, more recent pre-steady state evidence (9) from studies of S-1 binding and Tm activity indicates that the cooperative unit for Tm-actin may be as low as 5 actins instead of the 7 generally assumed and as high as 10-12 actins in the presence of actin-Tm-Tn. These studies also provide evidence for a sub-state of the thin filament in which S-1 weak binding is blocked at low Ca^{++}. There are several important conclusions from these studies. First, although Ca^{++} is a required co-factor, it cannot fully turn on the thin filament; full thin filament activation requires strong crossbridges. Second, myosin heads bind cooperatively; cooperative binding of myosin heads requires Tm, which alone can confer cooperativity; Tm and S-1 may compete for the same binding sites on actin. Third, apart from its role as the Ca^{++} receptor Tn appears to increases the size of the cooperative unit and regulate a blocked state of thin filament with regard to weak S-1 binding.

MODULATION OF THE ALLOSTERIC/COOPERATIVE ACTIVATION PROCESSES

With regard to cardiac energetics and thin filament activity, the following questions seem most relevant: What is likely to happen in the transition from the diastolic (relaxed) to the systolic (active) state? How might the steric and allosteric/cooperative processes be modified to alter the amount of Ca^{++} required to achieve a particular contractile state? Release of Ca^{++} into the myofilament space results in Ca^{++} binding to a single regulatory site on TnC; Ca^{++} acts at some distance from the S-1 binding site on actin (i.e. allosterically) to trigger contraction through a series of protein-protein interactions involving TnI, TnT and Tm. The trigger may involve two types of processes, a release from a blocked state and an increase in the rate constants for transition from the weak to the strong binding state. Tropomyosin or TnI, and potentially TnT (20), are moved from their blocking position in a few functional units by strong crossbridges, near neighbor crossbridges promote further activation of the thin filament and thus activation spreads and additional functional units are engaged in the contraction. Thus, a particular contractile state may be perceived as a balance between the allosteric activation of the myofilaments by Ca^{++} and the cooperative activation of the myofilaments by crossbridges. The force that is generated appears to be a linear function of the distribution of crossbridge states and force generated by each state.

The questions that follow are: Whether, when and how is the allosteric/cooperative mechanism modified physiologically? Is it modified in pathophysiologic states? Can the process be modified pharmacologically? The answer to these questions is the subject of investigation in many laboratories and a complete answer is beyond the scope of this chapter. Yet, it is important to point out that there is compelling evidence that alteration of the response of the myofilaments to Ca^{++} is regulated physiologically in association with intrinsic as well as extrinsic regulation of the heart; and further, that the response of the myofilaments to Ca^{++} is altered in common pathologies such as ischemia/acidosis, cardiac stunning, and genetically linked hypertrophic myopathies. In addition, there are diverse classes of pharmacological agents that appear to alter the allosteric/ cooperative process at the level of TnC, at the level of the cooperative activation of the myofilaments, and at the level of the actin-myosin reaction itself.

Physiological Modulation of the Response of the Myofilaments to Ca^{++}

There are substantial data indicating that length dependence of myofilament activation forms the basis of Starling's Law (1,27,28). Thus, myofilament response to Ca^{++} appears to be an important determinant of the shape of the relation between end-systolic volume and end-systolic pressure. There is also substantial evidence that adrenergic stimulation of the heart is associated with phosphorylation of the myofilaments by protein kinase A (PKA), protein kinase C (PKC) and also Ca^{++}-calmodulin dependent protein kinase, to name the most well studied examples (21,29). Importantly, covalent modulation of the myofilament proteins via these kinases affects their activity and their control by Ca^{++} (21,29). An important aspect of regulation by PKC pathways is the idea that phosphorylation may affect the extent to which the thin filament is turned on. Possible hypotheses for the effect of phosphorylation of TnI and TnT are as follows: 1) regulation of the blocked state, 2) modulation of allosteric effects on actin structure, 3) modulation of the energies of interaction among the proteins in a functional unit and perhaps between functional units of the thin filament.

Pathological and Pharmacological Modulation of the Response of the Myofilaments to Ca^{++}

There are alterations in the response of the myofilaments to Ca^{++} that occur during conditions associated with depressed cardiac function. One example is during acidosis, in which it has been shown that force generating capabilities of heart muscle are considerably depressed under conditions in which the amplitude of Ca^{++} transient during the twitch is, in fact, increased (30). Similar effects occur during the depression in cardiac function associated with hypoxia and accumulation of ADP or Pi, the products of ATP hydrolysis and the breakdown of creatine phosphate (1,27). There are also potential long term changes in the structure and function relations of the myofilaments associated with breakdown of myofilament proteins (31,32), long term changes in covalent modulation (21) or altered gene expression (33,34) that may affect the myofilament response to Ca^{++}. A tie between altered gene expression and heart failure is the identification of mutations in the myosin heavy chain gene that form the molecular basis of familial hypertrophic cardiomyopathy (34). More recent evidence links missense mutations in TnT and in Tm to this same disease, which is emerging as a "sarcomeric" disease (35). Reexpression of a fetal isoform of TnT in various forms of heart failure has also been correlated with the well known depression in maximum myofibrillar ATPase associated with the disease (33).

The complexity of myofilament regulation suggests a rich array of possible mechanisms by which myofilament response to Ca^{++} might be manipulated pharmacologically. Recent evidence indicates that alterations in the response of the myofilaments to Ca^{++} may indeed alter the energy requirements for pumping function of the heart. The inotropic agent Acardi (Pimobendan; UDCG 115BS) increases Ca^{++} binding to TnC in myofilaments and is also an inhibitor of cAMP phosphodiesterase (Type III) (36). The introduction of Acardi into clinical use in heart failure treatment in Japan shows early promise, but widespread utilization of this calcium sensitizer must await more extensive clinical trials. Convincing evidence that such agents may affect cardiac energetics in humans comes from recent studies showing that the compound MCI-154 (37) reduces myocardial oxygen consumption in humans. The proposed mechanism is that an altered response of the myofilaments to Ca^{++} reduces the cost of non-mechanical work (38). This fits well with the premise of this paper, which began with the idea that one way to alter the oxygen cost of non-mechanical work may be to alter the amount of Ca^{++} required to activate the myofilaments.

ACKNOWLEDGEMENTS

This author is supported by research grants NIH RO1-HL22231 and R01 HL49934.

REFERENCES

1. Solaro, R.J., Pan, B.S.: Control and modulation of contractile activity of cardiac myofilaments, in Sperelakis N (ed): *Physiology and Pathophysiology of the Heart.* Kluwer Academic Publishers, Boston, 291-293, 1988.

2. Namba, T., Takaki, M., Araki, J. et al. Ca^{++} sensitivity of contractile machinery and Ca^{++} handling energy. Simulation. Jap Heart J 34:601-616, 1993.

3. Lehrer, S. The regulatory switch of the muscle thin filament: Ca^{++} or myosin heads? J Mus Res Cell Motility 15:232-236, 1994.

4. Millar, N., Homsher, E. The effect of phosphate and Ca on force generation in glycerinated rabbit skeletal muscle fibers. J Biol Chem 265:20234-20240, 1990.

5. Kawai, M., Saeki, Y., Zhao, Y. Crossbridge scheme and the kinetic constants of elementary steps deduced from chemically skinned papillary and trabecular muscles of the ferret. Circ Res 73:35-50, 1993.

6. Pan, B.S., Solaro, R.J. Calcium-binding properties of troponin C in detergent-skinned heart muscle fibers. J Biol Chem 262:7839-7849, 1987.

7. McKillop, D.F.A., Geeves, M.A. Regulation of the interaction between actin and myosin subfragment 1: evidence for three states of the thin filament. Biophys J 65:693-701, 1993.

8. Brenner, B. Muscle mechanics and biochemical kinetics. In: Squire JM editor. Molecular Mechanisms in Muscular Contraction, MacMillan Press, London 77-149, 1990.

9. Geeves, M.A., Lehrer, S.S. Dynamics of the muscle thin filament regulatory switch: the size of the cooperative unit. Biophys J 67:273-282, 1994.

10. Kress, M., Huxley, H.E., Faruqi, A.R., Hendrix, J. Structural changes during activation of frog muscle studied by time resolved X-ray diffraction. J Mol Biol 188:325-342, 1986.

11. Lehman, W., Craig, R., Vibert, P. Nature 368:65-67, 1994.

12. Poole, K.P.V., Lorenz, M., Evans, G., Rosenbaum, G., Holmes, K.C. Biophys J 66A:347, 1994.

13. Popp, D., Maeda, Y. Calcium ions and the structure of muscle actin filament: an X-ray diffraction study. J Mol Biol 229:279-285, 1993.

14. Eaton, B.L., Kominz, D.R., Eisenberg, E. Correlation between the inhibition of the acto-heavy meromyosin ATPase and the binding of tropomyosin to F-actin: effects of Mg, KCl, troponin I and troponin C. Biochemistry 14:2718-2725, 1975.

15. Bremel, R., Murray, J., Weber, A. Manifestations of cooperative behavior in the regulated actin filament during actin-activated ATP hydrolysis in the presence of calcium. Cold Spring Harbor Symp Quant Biol 37:267-275, 1973.

16. Brenner, B. Changes in calcium sensitivity at the crossbridge level. In: Lee JA, Allen, D.G., editors. Modulation of cardiac calcium sensitivity, Oxford University Press, Oxford 197-214, 1993.

17. Walker, J.W., Lu, Z., Moss, R.L. Effects of Ca^{++} on the kinetics of phosphate release in skeletal muscle. J Biol Chem 267:2459-2466, 1992.

18. Kawai, M., Cox, R.N., Brandt, P.W. Biophys J 35:375-384, 1981.

19. Zot, A.S. and Potter, J.D.: Structural aspects of troponin-tropomyosin regulation of skeletal muscle contraction. Ann. Rev. Biophys. Biophys. Chem. 16:535-559, 1987.

20. Heeley, D.H., Smillie, L.B. Interaction of rabbit skeletal muscle troponin T and F-actin at physiological ionic strength. Biochemistry 27:8227-8231, 1988.

21. Solaro, R.J.: Protein phosphorylation and the cardiac myofilaments. In: Protein Phosphorylation in Heart (Solaro, R.J., ed.) CRC Press, Inc., Boca Raton, pp 129-156, 1986.

24. Williams, D., Greene, L., Eisenberg, E. Cooperative turning on of myosin subfragment 1 ATPase by the Tn-Tm-actin complex. Biochemistry 27:6987-6993, 1988.

25. Green, L.E., Eisenberg, E. Cooperative binding of myosin subfragment-1 to the actin-troponin-tropomyosin complex. Proc Nat Acad Sci USA 77:2616-2620, 1980.

26. Lehrer, S., Ishii, Y. Fluorescence properties of acrylodan-labeled tropomyosin and tropomyosin-actin: Evidence for myosin subfragment 1 induced changes in geometry between tropomyosin and actin. Biochemistry 27:5899-5906, 1988.

27. Moss, R.L. Ca^{++} regulation of mechanical properties of striated muscle: mechanistic studies using extraction and replacement of regulatory proteins. Circ Res 70:865-884, 1992.

28. Allen, D.G., Kentish, J.C.: The cellular basis of the length-tension relation in cardiac muscle. J Mol Cell Cardiol 17:821-840, 1985.

29. Solaro, R.J. Modulation of activation of cardiac myofilaments by beta-adrenergic agonists. In: Lee JA, Allen DG, editors. Modulation of cardiac calcium sensitivity, Oxford University Press, Oxford 161-177, 1993.

30. Solaro, R.J. Lee, J., Kentish, J. and Allen, D.A. Differences in the response of adult and neonatal heart muscle to acidosis. Circ. Res. 63:779-787, 1988.

31. Westfall, M.V., Solaro, R.J.: Alterations in myofibrillar function and protein profiles following global ischemia in rat hearts. Circ Res 70:302-313, 1992

32. Margossian, S.S., White, H.D., Caulfield, J.B., Norton, P., Taylor, S., Slayter, H.S.: Light chain 2 profile and activity of human ventricular myosin during dialed cardiomyopathy. Circulation 85:1720-1723,1992.

33. Anderson, P.A.W., Malouf, N.N., Oakeley, A., Pagani, E.D., Allen, P.D.: Troponin T isoform expression in humans: A comparison among normal and failing adult heart, fetal heart, and adult and fetal skeletal muscle. Circ Res 60:1226-1233, 1991.

34. Tanigawa, G., Jarcho, J.A., Kass, S., Solomon, S.D., Vosberg, J.G., Seidman, J.G., Seidman, C.E.: A molecular basis for familial hypertrophic cardiomyopathy: an α/β cardiac myosin heavy chain hybrid gene. Cell 622:991-998,1990.

35. Thierfelder, L., Watkins, H., MacRae, C. et al. Alpha-tropomyosin and cardiac troponin T mutations cause familial hypertrophic cardiomyopathy: a disease of the sarcomere. Cell 77:701-712, 1994.

36. Mori, M., Takeuchi, M., Takaoka, H., et al. New Ca^{++} sensitizer, MCI-154, reduces myocardial oxygen consumption for non-mechanical work in diseased human hearts. Circulation I217, 1994.

37. Fujino, K., Sperelakis, N., Solaro, R.J. Sensitization of dog and guinea pig cardiac myofilaments to Ca^{++}-activation and inotropic effect of pimobendan: Comparison with milrinone. Circ Res 63:911-922, 1988.

4. Explaining Load-dependent Ventricular Performance and Energetics Based on a Model of E-C Coupling

Daniel Burkhoff, Matthew Schnellbacher, Richard A. Stennett, Donna Zwas, Kazuhide Ogino and James P. Morgan

INTRODUCTION

The course of research in cardiovascular physiology has been influenced greatly by the *time-varying elastance [E(t)] theory of ventricular contraction* which was introduced in the early 1970's (1). In principle, the E(t) function describes the time variations in instantaneous ventricular volume-elastance during a beat and was proposed as a load-independent characterization of dynamic ventricular pump properties. Additionally, the value of E(t) at end systole (E_{es}), which is the slope of the end-systolic pressure-volume relationship (ESPVR), varies directly with inotropic background and it was therefore proposed that E_{es} could serve as a load-independent index of ventricular *contractility*. Evaluation of the energetics of the E(t) theory also led to the notion of a load-independent relationship between energy consumption (in the form of oxygen, mVO_2) and mechanical energy liberation indexed by the pressure-volume area (PVA) (2).

However, several shortcomings of this theory have been identified which have brought into question some of the fundamental assumptions of both the ESPVR and of E(t). It has been shown that, in general, the ESPVR is nonlinear and the degree of nonlinearity depends upon the inotropic state (3). Furthermore, the ESPVR appears to have a greater degree of load dependence, particularly when ejection patterns are provided by physiologic loading systems (2). Interestingly, the nature of the load dependence identified recently is such that the E_{es} can be greater on ejecting beats than on isovolumic beats (i.e., an apparent ejection-mediated increase in contractile state). Similarly, under loading conditions where ejection increases E_{es} over that observed during isovolumic conditions, the slope of the mVO_2-PVA relationship is decreased, suggesting an apparent increase in overall metabolic efficiency of contraction during ejection despite the increase in E_{es} (2). Finally, there are several aspects of the time course of contraction which have been shown to be highly load dependent (4) which are not accounted for by the E(t) theory which assumes a load-invariant time course of contraction.

Another limitation of the E(t) theory is that it provides a phenomenologic description of ventricular properties. While this is very useful in understanding the role of the heart in overall cardiovascular performance (5), there is increasing interest in being able to describe ventricular contractile properties in terms of the fundamental processes responsible for contraction. In fact, investigators have already attempted to explain some of the deviations of ventricular behavior from the E(t) theory based upon the notion that loading conditions can impact on the kinetic interactions between calcium, actin and myosin; a phenomenon broadly referred to as *length-dependence of activation* (6,7).

Recently, the feasibility of explaining a wide range of ventricular properties in terms of a 4-state biochemical scheme which accounts for the interactions between calcium, thin filaments and thick filaments and how load impacts on the kinetics of those interactions has been explored (8). This has resulted in the inception of a different way of thinking about ventricular contraction which will be summarized in this brief

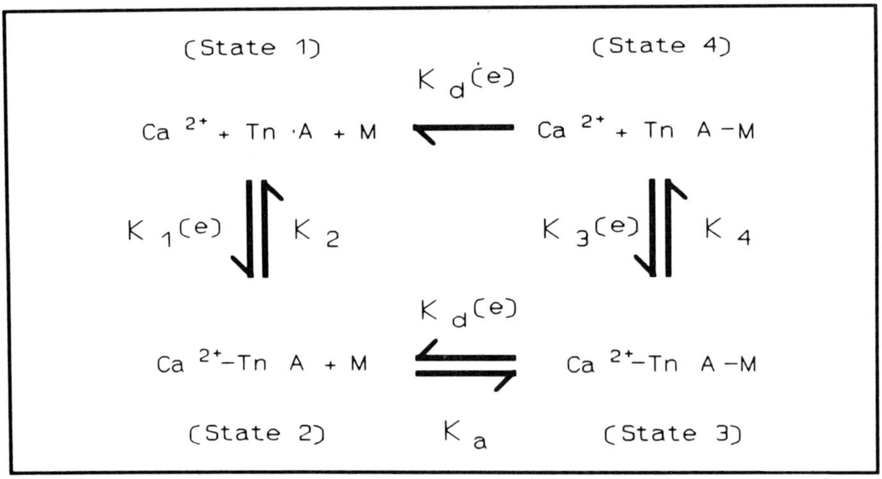

Fig. 1. 4-state biochemical scheme used to relate calcium transient to muscle force and, ultimately, left ventricular pressure generation. See text for further details.

review. In addition, some of the insights into the differences between isovolumic and ejecting contractions will be explored within the context of this new theory. Finally, preliminary experimental efforts to test the feasibility of the new theory will be reviewed.

METHODS

Theoretical Considerations

There is mounting evidence that a 4-state biochemical scheme originally proposed by Hill for skeletal muscle (9) (Fig. 1) may account for the temporal relations between the intracellular calcium transient $\{[Ca^{++}]_i(t)\}$ and force in cardiac muscle. This scheme provides one plausible means of accounting for the kinetic interactions between actin (***A***), myosin (***M***), troponin C (***Tn***) and calcium ($\boldsymbol{Ca^{++}}$) during a contraction cycle. It is possible to use this reaction scheme to predict the time course of muscle force (or stress, σ) generation from a specified free calcium transient under different loading conditions. To accomplish this, force is assumed to be proportional to the number of actin-myosin bonds (with proportionality constant α) which, in this formulation, can exist in two states: one with calcium bound (State 3), and one without calcium bound (State 4). Length (ϵ) dependence of activation was introduced in two steps of the model. First, since length has been shown to influence the calcium affinity of the myofilaments, length-dependence of rate constant values is introduced as symbolized by the ϵ associated with the various K's. Second, length affects the ability of the myofilaments to generate force for a given degree of activation, so that the proportionality constant between the number of cross-bridge units and force (α) was varied with ϵ. Finally, the well known *force-velocity relationship* (10) of muscle was introduced in a phenomenological manner by adding a term which decreases force in proportion to the velocity of shortening. These factors are summarized in the following equation:

$$\sigma(t,\epsilon)=\alpha(\epsilon)\{[\mathrm{Ca}^{++}\text{-}Tn\cdot A\text{-}M](t)+[Tn\cdot A\text{-}M](t)\}\text{-}\nu R \qquad (1)$$

where ν is the velocity of shortening ($\nu \equiv d\epsilon/dt$) and R is the proportionality constant. Further details of the underlying assumptions have been provided previously (8).

Computer Simulation

In order to test the feasibility of this theory, numerical techniques were used to determine whether the proposed model could account for the numerous phenomena observed in the ventricle which cannot be explained by the E(t) theory; these have been reviewed briefly above (see Introduction). The differential equations governing the 4-state scheme were programmed on a computer. The driving function for these equations was taken as the instantaneous concentration of free calcium $\{[Ca^{++}]_i(t)\}$ which was obtained from previously published data (11). Initial values for the rate constants were taken from the literature (12) and adjusted slightly in order to provide a good fit between the muscle stress measured simultaneously with $[Ca^{++}]_i(t)$ (11). Next, cardiac muscle with properties described by these differential equations and Eq. 1 were used to mathematically construct a ventricle via a simple thick walled spherical model. Diastolic chamber properties were assumed to be totally passive and were characterized by a simple relationship between pressure ($P_{passive}$) and volume (LVV): $P_{passive} = B\,(e^{A.LVV}\text{-}1)$. Total ventricular pressure was set equal to the sum of the $P_{passive}$ and the active pressure generated by muscle contraction. Thus, from the set of differential equations describing the system of Fig. 1 with length-dependent rate constants, the free calcium transient taken from the literature, assumed relationship between actin-myosin bonds, velocity of shortening and muscle stress (Eq. 1) and assumed spherical ventricular geometry, it is possible to define the instantaneous relationship between ventricular volume and pressure. Finally, the theoretical ventricle constructed in this manner was coupled to a realistic representation of arterial afterload impedance (i.e., the Windkessel model) so that the model ventricle can be loaded in a physiologic manner and previously performed experiments could be duplicated for the model ventricle.

Some of the phenomena to be examined deal with relationships between energy consumption and mechanical work done by the ventricle. In real hearts, energy consumption is indexed by myocardial oxygen consumption (mVo_2), since this relates closely to ATP consumption. ATP is used to support many different cellular processes, including basal metabolism, calcium cycling and force generation. The issue of main concern in the present study is in assessing the energy cost of force (pressure) generation at different loading conditions. It was assumed that basal metabolism and the energy for calcium cycling are independent of loading conditions (starting assumptions introduced to simplify the present analysis which may, in fact, be incorrect). The principal energy consuming process during contraction is the hydrolysis of ATP required to uncouple the actin-myosin bond. Therefore, energy consumption for force generation was assumed to be proportional to the number of actin-myosin bonds broken during the course of a contraction and was expressed in these arbitrary units. Mechanical work done by the ventricle was quantified, as has been done for real ventricles, by the *pressure-volume area* (PVA) which has been described in detail elsewhere (13).

The main goal of the simulation aspects of this study was to determine whether previously observed load and contractility dependent changes in end-systolic pressure, mechanical efficiency, time course of contraction and time course of relaxation could be explained by the relatively simple theory of muscle contraction presented above. Detailed sets of data have been obtained previously in isolated canine hearts which summarize these observations (2-4). For each of the phenomena investigated previously in isolated canine hearts, protocols were reproduced as closely as possible in the simulation and the analyses of the resulting data were identical to those performed previously on data obtained from the real hearts.

Measurement of $[Ca^{++}]_i(t)$ in Intact Rat Hearts

In order to provide an initial experimental test of the new theory, we set out to determine whether the proposed biochemical scheme could account for the temporal relations between the intracellular calcium transient, $[Ca^{++}]_i(t)$, and the time course of isovolumic pressure generation, LVP(t). We studied isovolumically contracting rat hearts in which $[Ca^{++}]_i(t)$ could be measured using a previously described aequorin macroinjection technique (14,15). Briefly, hearts were excised from anesthetized rats and perfused on a modified Langendorf perfusion apparatus with Krebs-Henseleit solution. A balloon was placed in the LV chamber through the mitral valve and balloon volume was adjusted for an end-diastolic pressure of ~10 mmHg. Between 3 and 5 μl of an aequorin solution (aequorin 1 mg/ml, NaCl 154 mM, KCl 5.4 mM, $MgCl_2$ 1 mM, HEPES 12 mM, glucose 11 mM and EDTA 0.1 mM) is injected just under the epimysium of the infero-apical region of the heart using a glass micropipette with an inner diameter of ~30μm. It has been shown that the aequorin diffuses into epicardial muscle cells with minimal damage to the cells near the injection site (as evidenced by light microscopy) (15). The heart and perfusion apparatus are placed inside a light tight box and lowered into a specially designed glass organ bath with a concavity at its base; the aequorin injection site is placed in contact with this base so that the aequorin luminescence is emitted through the bottom of the bath. In turn, the bottom of the organ bath is positioned at the focal point of an ellipsoidal light collector which directs the light to a photomultiplier tube (9235QA, Thorn EMI, Fairfield NJ).

The method of calibrating the light signal into an absolute concentration of calcium is the same as used in papillary muscles (15). At the end of the experiment, the heart is perfused with an 8mM calcium-5% Triton X-100 solution which lyses the cells and exposes the remaining aequorin to a high amounts of calcium (14). In order to calibrate a particular calcium transient, a quantity referred to as L_{max} must be determined which is the integral of the aequorin signal from the time of the chosen transient to the end of the experiment including the light collected during the lysis procedure. After determining the appropriate L_{max}, the particular light transient [L(t)] is converted to $[Ca^{++}]_i(t)$ using a previously determined calibration curve (15).

The measured $[Ca^{++}]_i(t)$ curves were then used as the driving functions for the differential equations of the 4-state model of contraction and rate constants were adjusted to try to fit the simultaneously measured LV pressure curves with the LV pressure curve predicted from the model. The degree of concordance between model predicted LV pressure [$LVP_p(t)$] and measured LV pressure [$LVP_m(t)$] curves was quantified by determining the root mean squared difference (D_{RMS}) between the two curves throughout the beat:

$$D_{RMS}=\frac{1}{T}\int\sqrt{[LVP_m(t)-LVP_p(t)]^2}dt \qquad (1)$$

where T is the contraction duration and the integration is performed over the entire heart beat.

RESULTS

The New Theory Explains Load-Dependence of Ventricular Contraction, Relaxation and Energetics

With cardiac muscle properties represented by the 4-state biochemical scheme, ventricular properties derived from these and a spherical geometrical transformation and the resulting model ventricle coupled to a Windkessel afterload, theory predictions were compared to actual ventricular behavior. A detailed review of this analysis has been presented previously (8) and only the important highlights will be reviewed here.

The first comparison between theory predictions and experimental results was the load dependence of the ESPVR (2). For the real isolated canine hearts, the parameters of the Windkessel afterload were set to provide a ventricular ejection fraction of approximately 50%. Pressure-volume loops were recorded at several different preload volumes. Then, the ventricle was constrained to contract isovolumically and pressure waves were recorded over the same range of ventricular volumes. From the resulting loops, the ESPVR was determined in the usual manner.

A comparison between ESPVRs measured under isovolumic and ejecting conditions obtained from an isolated heart is shown on the left side of Fig. 2; the solid line shows the ESPVR from the isovolumic conditions. As shown, the pressure-volume loops of the ejecting contractions break through the isovolumic ESPVR, thus comprising an ESPVR that falls well above that of the isovolumic contractions. As reviewed previously (2), the steady-state ESPVRs measured under ejecting conditions were typically shifted upwards compared to isovolumic contractions for ejection

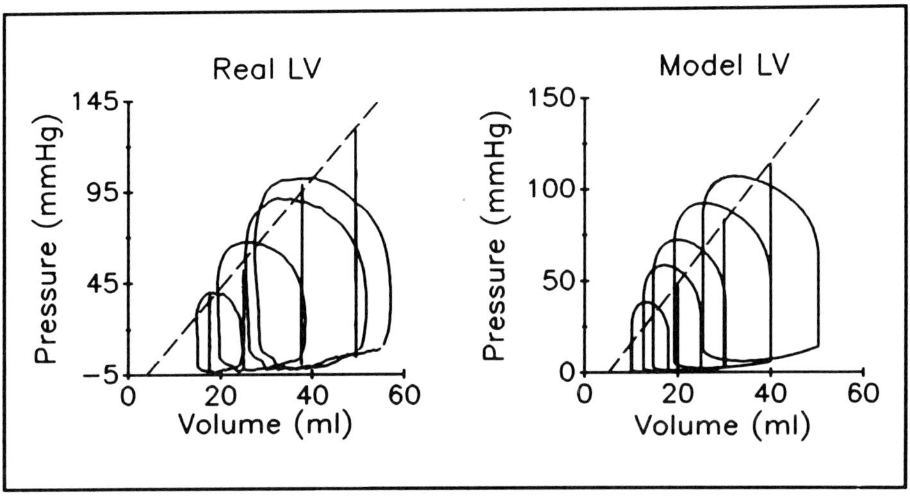

Fig. 2. Left: LV ESPVRs measured from isolated canine heart under ejecting and isovolumic conditions. Right: model ventricle behaves same way in response to the load change.

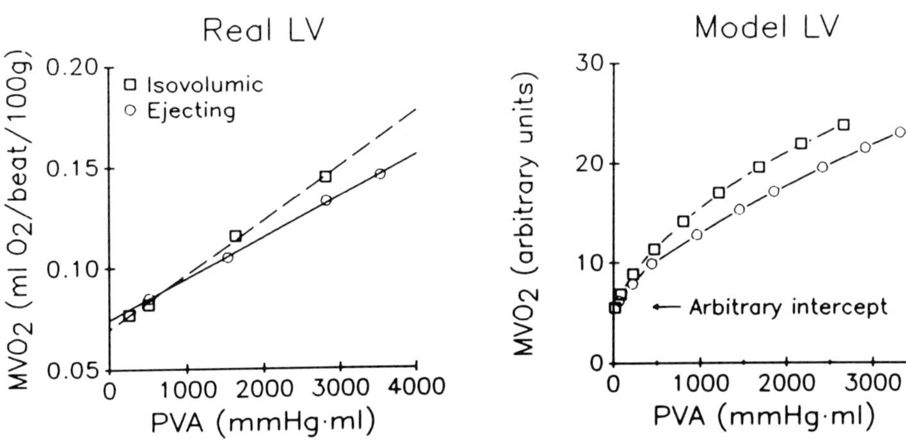

Fig. 3. Left: mVO_2-PVA relationship from isolated heart under isovolumic and ejecting conditions. Right: energetics of model LV responds the same way to the change in load.

fractions ranging between 20 and 60%. The behavior of the theoretical ventricle, shown on the right side of Fig. 2, exhibits the same property.

The next test of the theory was to examine the influence of ventricular volume on the characteristics of the isovolumic pressure waveform. Isovolumic pressure waveforms were recorded from real ventricles over a wide range of volumes (4). Each isovolumic pressure waveform was characterized by three parameters: T_{es}, time to end systole; τ, the time constant of relaxation; and D_{10}, an index of the overall duration of contraction, defined as the width of the pressure wave (in milliseconds) at a level 10% of the developed pressure.

Data from a real isolated ventricle revealed that each of these parameters increased as the volume of the isovolumic contraction was increased. The same protocol was executed in the theoretical ventricle and the results exhibit the same behavior.

The characteristics of the pressure waveform measured under isovolumic and ejecting conditions were then compared. In this protocol, pressure waveforms were measured under three conditions. First, the ventricle was allowed to eject against a specified afterload impedance; under these conditions, the end-diastolic and end-systolic volumes were noted. Next, the ventricle was forced to contract isovolumically at the end-diastolic volume of the ejecting beat. Third, the ventricle was forced to contract isovolumically at the end-systolic volume of the ejecting beat. The pressure waveforms measured under each condition were characterized by the parameters described above (T_{es}, τ and D_{10}). Data from the real ventricles (4) revealed a prolongation of T_{es}, a decrease in τ and relatively little variation in D_{10} on ejecting as compared to isovolumic beats. Again, the model ventricle exhibits very similar behavior.

Finally, the linearity of the ESPVR and how this linearity was altered by changes in contractile state was examined (3). For this protocol, the ventricle was constrained to contract isovolumically and the peak pressure was recorded at many volumes. For the real ventricle, contractile state was increased by infusion of Bay K 8644 (a calcium channel agonist) and decreased by infusion of nifedipine (a calcium channel antagonist). For the model ventricle, contractile state was increased by multiplying the calcium transient by a scaling factor of 1.5, and contractile state was decreased by multiplying the calcium transient by a scaling factor of 0.5. For the real ventricle, the ESPVR at baseline contractile state was linear. However, when contractile state was enhanced, the ESPVR became concave to the volume axis and it became convex to the volume axis when contractile state was decreased below baseline. The model ventricle exhibited the same behavior.

Interrelating Measured $[Ca^{++}]_i(t)$ and LVP(t) Using the New Theory

Aequorin light transients have been measured from six isolated, isovolumically contracting rat hearts. These light transients have been converted into intracellular calcium transients by measuring L_{max} and using a previously published calibration curve (15). The resulting calcium transients have been used as the driving function for the differential equations of the 4-state biochemical scheme and rate constants have been varied to provide the best fit between measured LV pressure and LV pressure predicted from the model. A typical example is shown in Fig. 4. Note that due to technical difficulties in determining the actual diastolic level of calcium in these experiments, this has been arbitrarily set to 0. In this case, the peak calcium concentration was 0.89 μM over the diastolic level. The peak of the calcium transient occurred very near to the time of LV dP/dt_{max}. In fact, analysis of several beats from all hearts studied indicated that the time difference between peak calcium and LV dP/dt_{max} was 13 ± 16ms (NS from 0); this is similar to the finding in isolated muscle strips that peak calcium occurs near the time of peak rate of rise of force (dF/dt_{max}). As shown, there is a very good agreement between measured and predicted pressure waves with the only noticeable deviations occurring during relaxation. D_{RMS} for this example was 0.24 mmHg, which means taken over the entire beat, there was only an average of 0.24 mmHg difference between measured and predicted curves. Taken from the 6 hearts as a whole, the mean ($\pm$SD) D_{RMS} was 0.25 ± 14 mmHg. These data indicate that not only can the 4-state model qualitatively account for a wide range of phenomena observed in intact canine hearts, it can be used to quantitatively relate calcium transients with measured ventricular pressure curves.

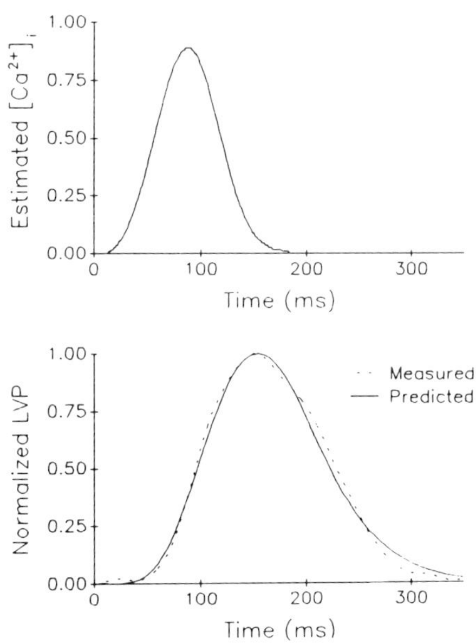

Fig. 4. Top: Calcium transient estimated from aequorin luminescence in isolated rate heart. Bottom: LV pressure predicted from the model is very close to that measured from the rat heart.

DISCUSSION

The time-varying elastance theory [*E(t)*] has provided a valuable way of studying ventricular pump properties and the role of the heart in the circulatory system.

However, when subjected to rigorous examination, many of the fundamental axioms of the E(t) theory do not hold for physiologically loaded isolated canine hearts or for hearts *in vivo*. The points of departure of ventricular behavior from E(t), namely load dependence of contraction, relaxation and energetics, have been explained in principle by an alternate theory. The theory is based upon a 4-state biochemical scheme that models the interactions between calcium and the myofilaments. This scheme was initially proposed by Hill to explain behavior of skeletal muscle (9). Yue invoked this scheme to explain the temporal relations within a beat between free calcium concentration (estimated from aequorin luminescence) and cardiac muscle force generation (11). Further experimental evidence to support the validity of the scheme was provided by Peterson *et al* (12) who used a bioassay to estimate the time course of calcium bound to the myofilaments. This original scheme was supplemented by incorporating two additional basic properties of cardiac muscle: length dependence of calcium binding affinity and load dependence of force generation. The biochemical scheme was used to provide the force-length-time properties of cardiac muscle which were used to construct a ventricle via a simple spherical geometry.

This theory was able to account for many fundamental aspects of ventricular performance. Most important of these are, as shown above: 1) enhancement of contractile strength (defined by the ESPVR) on ejecting compared to isovolumic beats; 2) improved contractile efficiency (defined by a decrease in the slope of the mVo_2-PVA relationship) on ejecting as compared to isovolumic beats; 3) appropriate volume-dependent changes in time to peak pressure, time constant of relaxation and duration of contraction on isovolumic beats and 4) ejection mediated prolongation of the time to end systole and speeding of relaxation.

While results obtained from the real and theoretical ventricles were qualitatively very similar in many regards, it is to be emphasized that these results are derived from model analysis and are thus subject to the limitations of the underlying assumptions built into the theory. Indeed, each assumption must be tested individually in future studies.

As a first step in this direction, calcium transients and pressure waves were measured from isovolumically contracting rat hearts. It has been demonstrated that when the measured calcium transient is used as the driving function for the set of differential equations describing the proposed system (Fig. 1), rate constants can be adjusted to provide a very good fit between measured LV pressure and pressure predicted from the theory. Interestingly, a separate analysis (not presented above) has shown that the concordance between measured and predicted pressure is significantly better with the 4-state scheme used (Fig. 1, with 7 adjustable rate constants) than with a 3-state model in which State 4 is omitted (with either 4 or 5 rate constants). This finding, which is similar to findings obtained from quantitative analyses of force and calcium transients, suggests that inclusion of a calcium-free force-generating state is important for explaining the dynamics of ventricular contraction and relaxation.

ACKNOWLEDGEMENTS

This work was supported by grants from NIH (1-R29-HL51885-01) and the Whitaker Foundation. D. Burkhoff was supported by an Investigatorship Award from the American Heart Association, New York City Affiliate, Inc.

REFERENCES

1. Suga H., Sagawa K. Instantaneous pressure-volume relationships and their ratio in the excised, supported canine left ventricle. Circ. Res. 35: 117-126. 1974.

2. Burkhoff D. et al. Contractile strength and mechanical efficiency of left ventricle are enhanced by physiological overload. Am. J. Physiol. 60 (Heart. Circ. Physiol. 29): H569-H578, 1991.

3. Burkhoff D. et al. Contractility-dependent curvilinearity of end-systolic pressure-volume relations. Am. J. Physiol. 252 (Heart. Circ. Physiol. 21): H1218-H1227, 1987.

4. Burkhoff D. et al. Impact of ejection on the magnitude and time course of ventricular pressure generating capacity. Am. J. Physiol. 265 (Heart. Circ. Physiol. 34):H899-H909, 1993.

5. Sagawa K. et al. Cardiac Contraction and the Pressure-Volume Relationship. Oxford University Press, Oxford 1988.

6. Jewell B.R. Review: A reexamination of the influence of muscle length on myocardial performance. Circ. Res. 40:221-230, 1977.

7. Allen D.G., Kentish J.C. The cellular basis of the length-tension relation in cardiac muscle. J. Mol. Cell. Cardiol. 17:821-840, 1985.

8. Burkhoff D. Explaining load dependence of ventricular contractile properties with a model of excitation-contraction coupling. J. Mol. Cell. Cardiol. 1994; in press.

9. Hill T. Two elementary models for the regulation of skeletal muscle contraction by calcium. Biophys. J. 44:383-396, 1983.

10. Sonnenblick E.H. Implications of muscle mechanics in the heart. Fed. Proc. 21:975-993, 1962.

11. Yue D.T. Intracellular $[Ca^{++}]$ related to rate of force development in twitch contraction of heart. Am. J. Physiol. 252 (Heart. Circ. Physiol. 21): H760-H770, 1987.

12. Peterson J.N. et al. Estimated time course of $Ca(^{++})$ bound to troponin C during relaxation in isolated cardiac muscle. Am. J. Physiol. 260 (Heart. Circ. Physiol. 29):H1013-H1024, 1991.

13. Suga H. Ventricular Energetics. Phys. Rev. 70:247-277, 1990.

14. Bentivegna L.A. et al. Altered calcium handling in left ventricular pressure-overload hypertrophy as detected with aequorin in the isolated, perfused ferret heart. Circ. Res. 69:1538-1545, 1991.

15. Kihara Y. et al. Direct measurement of changes in intracellular calcium transients during hypoxia, ischemia, and reperfusion of the intact mammalian heart. Circ. Res. 65:1029-1044, 1989.

5. Cardiac Contractions, PVA and Energetic Considerations Determined from a Cardiac Muscle Crossbridge Model

Tad W. Taylor, Yoichi Goto and Hiroyuki Suga

INTRODUCTION

As computers become more and more powerful, mathematical modeling and computer simulation becomes increasingly more important. Some recent personal computer designs offer performance that was in the realm of super computers only a decade ago. This allows chemistry and physics to be implemented in the mathematical modeling solution of biochemical problems. With sophisticated computer solutions available, various biochemical hypotheses can be quickly investigated, and this can aid in the experimental design and fundamental understanding of biochemical phenomena.

Just as the pressure-volume area (PVA) has been shown to linearly correlate with cardiac oxygen consumption, the force-length area (FLA) and also the tension-area-area have analogously been shown to correlate linearly with oxygen consumption (1-3). The FLA is analogous to PVA and corresponds to the area bounded by the end-systolic force-length relation, end-diastolic force-length relation, and the systolic force-length loop segment. To compute the FLA, cardiac twitch contractions were simulated by Huxley's sliding filament crossbridge muscle model coupled with parallel and series elastic components (4-7). The energetics of the contraction were based on ATP hydrolysis for the crossbridge cycling. The FLA as a measure of the total mechanical energy was computed for both isometric and isotonic contractions in a manner similar to the PVA (8-12). PVA correlates linearly with cardiac oxygen consumption, and since FLA is analogous to PVA, FLA should correlate with the ATP expended.

Yanagida et al. (13) experimentally determined that more than one crossbridge cycle occurred per ATP, typically six or more, and that the length of the crossbridge movement could be increased under an unloaded condition to 600 A. Recently, Lombardi et al. (14) reported that in the contractile response during lengthening of frog muscle fibers, there is a fast detachment of the crossbridges with a fast reattachment possible without an ATP hydrolysis occurring. This implies that the ATP hydrolysis can occur over several crossbridge cycles, as was hypothesized by Yanagida et al. (13).

Simulations comparing FLA with crossbridge cycling ATP usage showed that at lower muscle fiber activation levels (shorter initial fiber lengths and lower preload levels) FLA decreased more rapidly than the number of muscle fiber crossbridge cycles in both isometric and isotonic contraction cases. This suggests that one ATP can cause more than one crossbridge cycle at lower activation levels. If the number of crossbridge cycles to ATP ratio is allowed to increase at lower activation levels, Huxley's model is compatible with the experimental findings for FLA and PVA (8).

Using our 3-element cardiac muscle fiber model, simulations were done to see if it was possible to predict the cardiac Fenn effect (9). The Fenn effect was originally described in frog sartorius muscle in that a shortening skeletal muscle releases more energy than an isometric contraction (15-17). The Fenn effect has been explained as an increase in ATP splitting during isotonic contractions relative to isometric contractions (18,19), and has been recognized as a manifestation of the autoregulatory capability of muscle energetics in response to mechanics (20,21). Cardiac muscles share Mommaerts' unifying concept of the Fenn effect in that isotonic contractions involve liberation of energy in excess of the isometric contraction at a given force (22,23); this energy is equivalent to work.

The force-length energy (FLE, proportional to FLA) was computed in both isometric and isotonic contractions as a function of muscle fiber length (preload) in the isometric case and afterload in the isotonic contraction case. Simulation results demonstrated that the isotonic contractions produced a greater FLE than isometric contractions at every corresponding afterload, with the difference being equal to the work produced in the isotonic case, which is characteristic of the Fenn effect. The maximum energy utilization was observed at maximum force isometric contractions as has been experimentally observed in cardiac muscle. Changing the stiffness of the series elastic component did not change the Fenn effect behavior. Fenn-effect plots using crossbridge energy predictions showed behavior similar to the FLE plots, but the FLE to crossbridge energy ratio declined with decreasing force even though the efficiency has been experimentally found to be constant.

Several investigators have experimentally found that the force-time integral (FTI) varies non-linearly with energy expenditure over the course of a cardiac contraction. Crossbridge (CB) energy utilization was determined by considering the ATP hydrolysis for crossbridge cycling, and this CB energy was compared with the FLE in a contraction (10). This CB energy was calculated in both isometric and isotonic contractions as a function of contraction time and compared to the FTI. Simulation results demonstrated that the ratio of the FTI to CB energy varies strongly throughout the cardiac cycle for both isometric and isotonic cases, as has been observed experimentally. Simulations also showed that using the FLE component of energy versus the CB energy gave a better correlation between the total energetic predictions and the FTI, agreeing with recent findings that the crossbridge cycle to ATP hydrolysis ratio may not be coupled one-to-one, especially at lower force levels.

METHODS

As shown in Figure 1, a Maxwell-type three-component Hill model was used to represent the papillary muscle. The active muscle consists of a CE, which is freely extensible at rest, a SE in series with the CE, and finally a PE which gives the resting elasticity. It has been shown (5-7) that the PE and SE behavior can be modeled by exponential relations

$$P_{pe} = P_o[\exp\{K_p(L-L_o)\}-1] \qquad (1)$$

and

$$P_{se} = P_l[\exp(K_s L_{se})-1] \qquad (2)$$

where P_o, K_p and K_s are constants, P_l is the preload applied to the fiber, L_o is the initial fiber length, L is the fiber length at any instantaneous time, t, and L_{se} is the SE length at time t.

Huxley (4) proposed the sliding filament muscle sarcomere model which is used to simulate the CE in this study. Huxley postulated that tension is generated by the interaction of actin and myosin molecules, the interaction being generated by the myosin crossbridge. The reaction between actin and myosin can be represented by

$$A + M \underset{g}{\overset{f}{\rightleftharpoons}} A - M \qquad (3)$$

where A-M represents the attached crossbridges, A and M are actin and myosin, f is the association rate constant, related to the probability that a myosin crossbridge will bind to an actin molecule, and g represents the dissociation rate constant, related to the probability that an attached crossbridge will become detached. The equations for f(x) and g(x) are

$$x < 0,\ f(x) = 0,\ g(x) = g_2 \qquad (4)$$

$$0 < x < h,\ f(x) = f_1 x/h,\ g(x) = g_1 x/h \qquad (5)$$

$$x > h,\ f(x) = 0,\ g(x) = g_1 x/h \qquad (6)$$

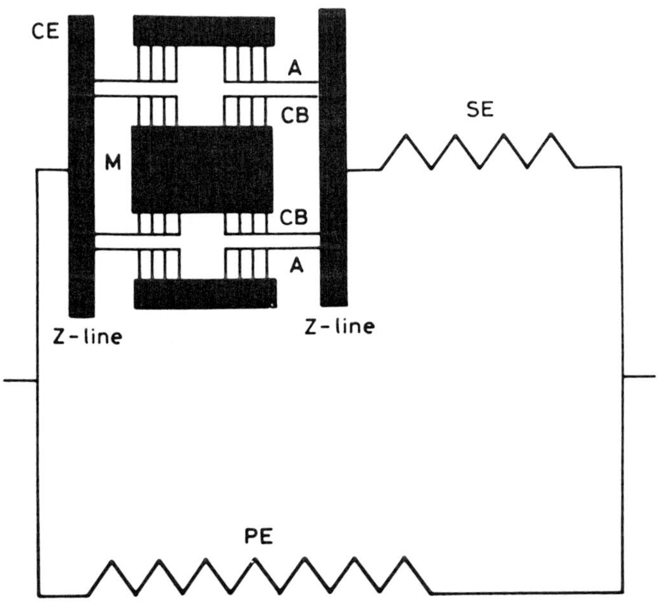

Fig. 1. Cardiac muscle fiber of the three-component Hill, or Maxwell-type model. A represents an actin filament, CB represents a crossbridge, and M represents a myosin filament in the CE. CE is the contractile element, PE is the parallel elastic element, and SE is the series elastic element.

In cardiac and skeletal muscles, tropomyosin exerts an inhibitory effect on the actin-myosin reaction. When calcium is released around the myofilaments following electrical stimulation, calcium reacts with troponin and allows the actin-myosin reaction to proceed. With n(x,t) being the number of attached cross-bridges in a sarcomere at position x and time t, it can be expressed as the partial differential equation (8-12)

$$dn(x,t)/dt - v(t)dn(x,t)/dx = [A(t)-n(x,t)]f(x) - n(x,t)g(x) \qquad (7)$$

where A(t) is the instantaneous number of actin molecules available to react with myosin molecules and v(t) is the velocity of the contractile element. Calcium is considered to react with the actin-troponin complex with one calcium ion involved

$$A\text{-}T + Ca^{++} \underset{c_2}{\overset{c_1}{\rightleftharpoons}} A + CaT \qquad (8)$$

where A-T is the concentration of actin that is inhibited by the tropomyosin complex, and c_1 and c_2 are the binding and dissociation rate constants. Other relations were implemented for the time varying calcium concentration, the contractile element force, the average number of crossbridge cycles, and the force-time integral (8-12).

RESULTS AND DISCUSSION

Crossbridge Cycle to ATP Hydrolysis Ratio

Figure 2 shows the relative normalized CB cycles to FLE ratios in the isometric case, where one represents the ratio at a fiber length of 9.2 mm. On each curve, the final value (at 9.2 mm fiber length) of the CB to FLE ratio was used to normalize all other values on the same curve. The standard case (Curve #1) reflects the CB to FLE ratio when the initial muscle fiber length is decreased; this causes the fiber force also to decrease. This curve implies that at smaller preload values Huxley's model predicts that more crossbridge energy is used than is actually observed by the FLE computations. The results in Figure 2 were done assuming one uniform cardiac muscle fiber is representative of the contraction occurring at the same time. The other 2 curves in Figure 2 present results from using different methods of varying the crossbridge to FLE ratio. In Curve #2, the crossbridge cycle rate was multiplied by the deactivation factor which is the relative time-independent amount of muscle fiber force deactivation that has occurred (8). This is analogous to decreasing the ATP to CB cycle ratio at shorter fiber lengths. This had the effect of keeping the normalized CB to FLE ratio close to unity. In Curve #3 in Figure 2, the crossbridge reach, h, was increased at lower activation levels by dividing h by the deactivation factor (8). This effect keeps the ratio closer to unity, as does multiplying the CB energy by the deactivation factor. This parallels Yanagida et al.'s (13) reasoning that either the crossbridge cycle to ATP ratio is increasing, or the possible length a crossbridge can reach has increased. Both effects will cause the crossbridge energy to FLE ratio to become more constant at lower deactivation levels.

Both decreasing the ATP to crossbridge cycle ratio and increasing the crossbridge reach are empirical measures, but they do allow better agreement between the cross-bridge energy and FLE. This lends support to the idea of loose coupling between mechanical and biochemical events.

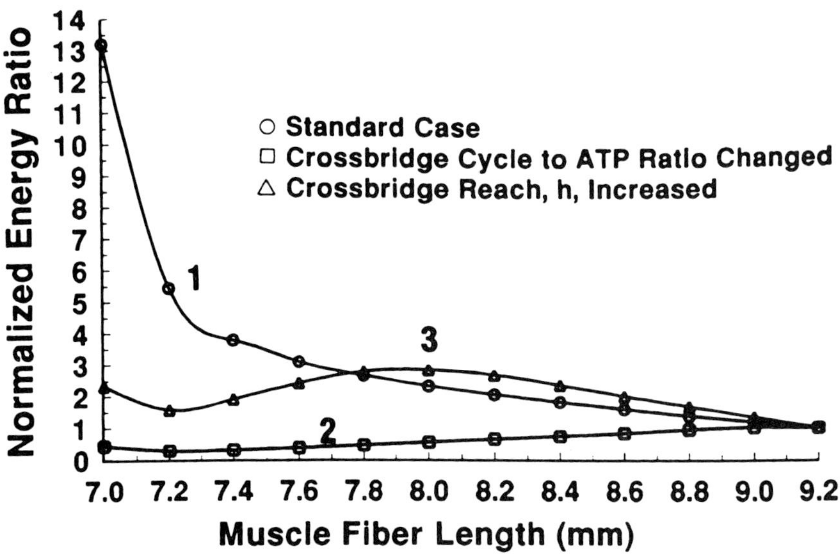

Fig. 2. Relative normalized CB cycles to FLE ratios in the isometric case, where one represents the ratio at a fiber length of 9.2 mm. For each curve, the final value (at 9.2 mm fiber length) of the CB to FLE ratio was used to normalize all other values on the same curve. Curve #1. The standard case showing the CB to FLE ratio when the initial muscle fiber length is decreased. Curve #2. The crossbridge cycle rate was multiplied by the deactivation factor. Curve #3. The reach of the crossbridge was changed by increasing the crossbridge reach by the inverse of the deactivation factor.

Yasumura et al. (24) showed that the oxygen consumption of an almost unloaded contraction from a relatively large end-diastolic volume only slightly exceeded the oxygen consumption of an isovolumic contraction. They interpreted this to imply that the end-diastolic volume does not significantly affect the oxygen consumption. This implies that at low cardiac muscle fiber force levels little oxygen and consequently little energy is used; hence several crossbridge cycles per ATP hydrolysis may be occurring.

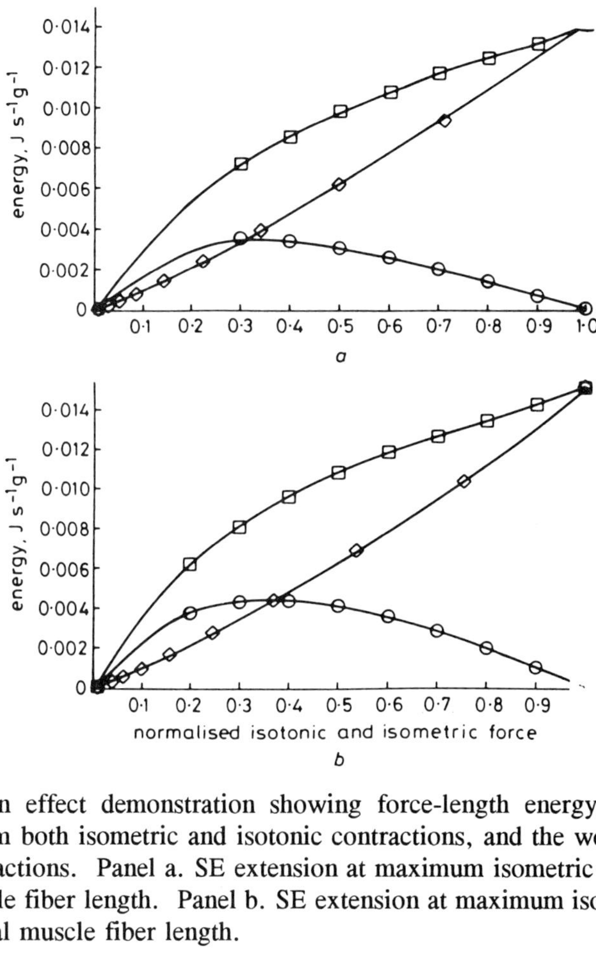

Fig. 3. Fenn effect demonstration showing force-length energy (FLE) values computed from both isometric and isotonic contractions, and the work obtained in isotonic contractions. Panel a. SE extension at maximum isometric force is 3% of the total muscle fiber length. Panel b. SE extension at maximum isometric force is 8% of the total muscle fiber length.

Fenn Effect

Figure 3 demonstrates the Fenn effect showing force-length energy values computed from both isometric and isotonic contractions, and also showing the work obtained in isotonic contractions. Panel a was constructed with an SE extension at maximum isometric force of 3% of the total muscle fiber length. To test the effect of the SE component extension value used, Panel b was made with an SE extension at maximum isometric force being 8% of the total muscle fiber length. The two panels resemble each other, with the work being larger with a larger SE extension in Panel b because the fiber contracts further. In Panel a, the maximum work occurs at approximately 0.3 relative force units, while in Panel b the maximum work occurs at approximately 0.35 maximum force units. The model is constructed so that in the isotonic case, the afterload must be greater than or equal to the PE force at the initial fiber length.

Both Panels a and b in Figure 3 demonstrate that the isotonic FLE is greater than the isometric FLE by a factor equal to the work performed, and the maximum FLE occurs at the maximum force in which case the isotonic and isometric FLE lines intersect. This is characteristic of the cardiac Fenn effect (22,23).

A different method to construct the Fenn effect diagram is presented in Figure 4, Panel a. Here the crossbridge energy is used for the isometric and isotonic contractions, and the SE extension at maximum isometric force is 3% of the total muscle fiber length. Figure 4, Panel a differs from Figure 3, Panel a in that in the isotonic case the maximum force does not predict the maximum crossbridge energy utilization, and the isometric CB energy line is concave downward versus being concave upward for the isometric FLE line shown in Figure 3, Panel a. The work should be the difference between the isotonic and isometric energy lines, and the work computed agrees with the results of Figure 3 but not with the results of Figure 4, Panel a.

Suga (23) suggested that the cardiac Fenn effect, with a larger energy consumption in the isotonic case at all comparable force levels, is not necessarily the result of a regulatory mechanism. This can be observed from the FLA diagram. In the isometric case, the FLA at a given force is just the area bounded by a vertical line at that force, the total force line, and the PE force line. In the isotonic case,the FLA at the same force consists of these components but there is also work area to the right of the isometric FLA area. This is clearly demonstrated in Figure 3, where the difference between the isotonic and isometric FLE is the work energy.

Experimental results have shown that the isometric energy curve is concave upward (25,26), while the isotonic force line is concave downward, as shown in Figure 3.

This contradicts the results of Figure 4, Panel a where the isometric CB energy curve is concave downward. This implies that the CB energy predictions are suspect, which again is likely due to the crossbridge cycle to ATP hydrolysis ratio not being coupled one-to-one.

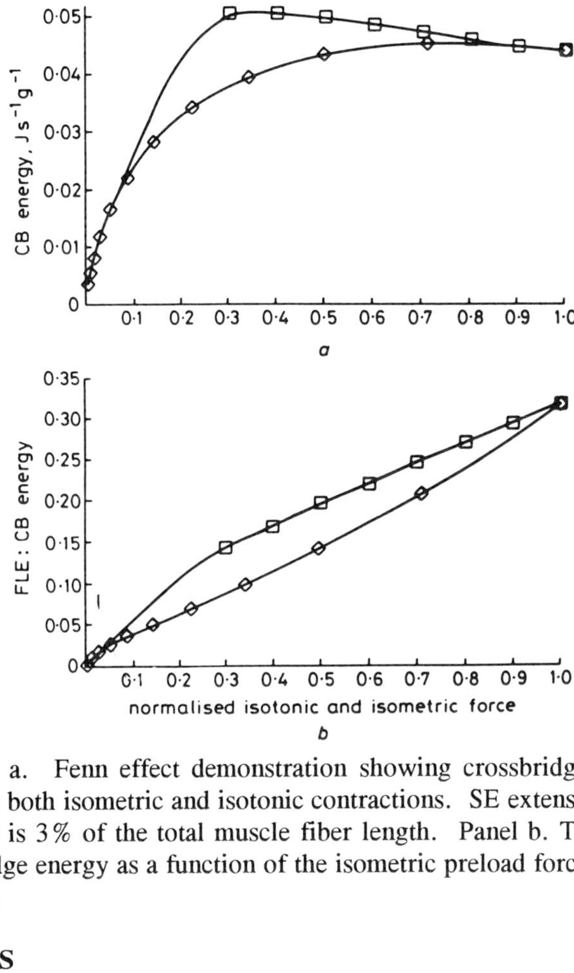

Fig. 4. Panel a. Fenn effect demonstration showing crossbridge energy values computed from both isometric and isotonic contractions. SE extension at maximum isometric force is 3% of the total muscle fiber length. Panel b. The ratio of FLE to the crossbridge energy as a function of the isometric preload force or the isotonic afterload force.

REFERENCES

1. Suga, H. Ventricular energetics. Physiological Reviews, 70:247-277, 1990.

2. Hisano, G., G. Cooper IV. Correlation of force-length area with oxygen consumption in ferret papillary muscle. Circ Res 61:318-328, 1987.

3. Goto, Y., S. Futaki, Y. Ohgoshi, H. Yaku and H. Suga. A new measure of left ventricular regional oxygen consumption: Systolic "tension-area". Circulation 80 (Supplement II):154, 1989.

4. Huxley, A.F. Muscle structure and theories of contraction. Progr Biophys Biophys Chem 7:255-318, 1957.

5. Wong, A.Y.K. Mechanics of cardiac muscle, based on Huxley's Model: mathematical simulation of isometric contraction. J Biochem 4:529-540, 1971.

6. Wong, A.Y.K. Mechanics of cardiac muscle, based on Huxley's Model: simulation of active state and force-velocity relation. J Biochem 5:107-117, 1972.

7. Panerai, R.B. A model of cardiac muscle mechanics and energetics. J Biochem 13:929-940, 1980.

8. Taylor, T.W., Y. Goto and H. Suga. Variable crossbridge cycling-ATP coupling accounts for cardiac mechanoenergetics. Am J Physiol 264:H994-H1004, 1993.

9. Taylor, T.W., Y. Goto and H. Suga. Myocardial mechanics and the Fenn effect determined from a cardiac muscle crossbridge model. Med & Biol Eng & Comput 31:377-381, 1993.

10. Taylor, T.W. et al. Comparison of the cardiac force-time integral with energetics using a cardiac muscle model. J Biochem 26:1217-1225, 1993.

11. Taylor, T.W., Y. Goto and H. Suga. On the solutions of Huxley-type models in cardiac muscle fiber contractions. J Theor Biol 165:409-416, 1993.

12. Taylor, T.W., Y. Goto and H. Suga. Cardiac muscle fiber force versus length determined by a cardiac muscle crossbridge model. Heart Vessels 7:200-205,1992.

13. Yanagida, T., T. Arata and F. Oosawa. Sliding distance of actin filament induced by a myosin crossbridge during one ATP hydrolysis cycle. Nature 316:366-369, 1985.

14. Lombardi, V., G. Piazzesi and M. Linari. Rapid regeneration of the actin-myosin power stroke in contracting muscle. Nature 355:638-641, 1992.

15. Fenn, W.O. A quantitative comparison between the energy liberated and the work performed by the isolated sartorius muscle. J Physiol 58:175-203, 1923.

16. Fenn, W.O. The relation between the work performed and the energy liberated in muscle contraction. J Physiol 58:373-359, 1924.

17. Woledge, E. Energetic aspects of muscle contraction. Academic Press, London. Soc Monogr 41, 1985.

18. Coleman, N.H., E.H. Sonnenblick and E. Braunwald. Myocardial oxygen consumption associated with external work: the Fenn effect. Am J Physiol 217:291-296, 1969.

19. Rall, J.A. Sense and nonsense about the Fenn effect. Am J Physiol 242:H1-H6,1982.

20. Homsher, E., W.F.H.M. Mommaerts and N.V. Ricchiuti. Energetics of shortening in twitches and tetanic contractions. II. Force-determined shortening heat. J Gen Physiol 62:677-692, 1973.

21. Mommaerts, W.F.H.M. What is the Fenn-effect? Muscle is a regulatory engine the energy output of which is governed by the load. Naturwissenschaft 57:326-330, 1970.

22. Suga, H. Energetics of the time-varying elastance model, a visco-elastic model, matches Mommaerts' unifying concept of the Fenn effect of muscle. Jap Heart J 3:341-353, 1990.

23. Suga, H. Variable series elasticity accounts for Fenn effects of skeletal and cardiac muscles. Am J Physiol 258:R457-R466, 1990.

24. Yasumura Y., et al. Minor preload dependence of O_2 consumption of unloaded contraction in dog heart. Am J Physiol 256:H1289-H1294, 1989.

25. Gibbs, C.L. Cardiac energetics and the Fenn effect. In Cardiac Energetics: Basic Mechanism and Clinical Implications. Edited by Jacob, R., Just, J., and Holubarsch, C., New York, Springer Publishing Co., Inc., 61-68, 1987.

26. Nozawa, T., Yasumura, Y., Futaki, S., Tanaka, N. and Suga. The linear relation between oxygen consumption and pressure-volume area can be reconciled with the Fenn effect in dog left ventricle. Circ Res 65:1380-1389, 1989.

6. Normalization of Emax and PVA

Motoaki Sugawara, Yukiyoshi Kondoh, Kiyoharu Nakano

INTRODUCTION

Analysis of the function of the whole ventricle is carried out conventionally by using the pressure-volume relation. However, the indexes derived from this relation do not represent the inherent properties of the myocardium; in other words, these indexes are not independent of the mass or geometrical features of the ventricle. For example, the value of Emax decreases with an increase in ventricular size even if myocardial contractility is assumed to be the same(1). If the elastic constants of the ventricle, such as Emax, were derived from the stress-strain system of the ventricular wall, they would be independent of the mass and shape of the ventricle, and represent local properties of the myocardium of the ventricular wall. However, there are presently no methods to measure stresses directly, and there are considerable discrepancies among the stress distributions in the ventricular wall calculated using various simplifying assumptions for the ventricular geometry and constitutive equations. Therefore, detailed stress-strain analyses are, at least clinically, of little practical utility.

Several methods of normalizing Emax have been proposed, but these corrections still remain controversial. It is basically difficult to normalize Emax properly, since the definition of Emax includes the ventricular volume, which is the cavity volume, and it is impossible to express the strain of the ventricular wall in terms of the cavity volume alone. Two ventricles with the same end-diastolic volume may have different unstressed volumes. Two ventricles with the same unstressed volume may have different wall thicknesses. Under such conditions, the same amount of volume change does not necessarily give the same amount of strain change to the myocardium of the ventricular wall.

We have developed methods of analyzing regional myocardial function of the ventricle based on the relation between the mean wall stress σ and the natural logarithm of reciprocal of wall thickness ln(1/H)(2-6). The mean wall stress is the average of the circumferential and meridional components of the wall stress. According to the incompressibility of the myocardium, ln(1/H) is proportional to the areal strain of the midwall layer of the ventricular wall(3-6). Therefore, the σ -ln(1/H) relation is one of the simplest forms of the stress-strain relation of the myocardium of the ventricular wall.

We have reported the following experimental and clinical results: 1) The area surrounded by the loop described by the σ -ln(1/H) relation during a cardiac cycle gives regional stroke work per unit volume of myocardium(3,4). 2) The end-systolic σ -ln(1/H) data point moves on the same single line during pressure and volume loadings, and its slope is a contractility index independent of ventricular size(1,5,6). Based on these results, we propose theoretically that the σ -ln(1/H) relation and the end-systolic σ -ln(1/H) relation are the normalized equivalents of the pressure-volume relation and Emax.

WHY DOES EMAX DECREASE WITH VENTRICULAR VOLUME?

Figure 1 shows the relation between Emax and end-diastolic volume in 25 normal dogs during β -blockade to stabilize the inotropic state. The end-diastolic volume of these dogs ranged from 14 to 82 ml. Emax decreases with ventricular size by a power function. Now let us consider why Emax decreases with ventricular size.

Based on the general theory of solid mechanics, we use the assumption that intrinsic elastic constants of the ventricular myocardium, which are independent of load and geometry, should be defined by stress-strain relations. Consider ventricles 1 and 2 of differing size but with the same Emax (Fig.2). The end-systolic pressure-volume relations of the two ventricles are given by

$$P_1 = \mathrm{Emax}(V_1 - V_{01})$$
$$P_2 = \mathrm{Emax}(V_2 - V_{02}),$$

where V_{01} and V_{02} are the volume intercepts of the Emax lines.Let us consider the same increment of end-systolic volume, ΔV, of ventricles 1 and 2. This volume increment, ΔV, causes the same increment of end-systolic pressure, ΔP, for both the ventricles, since the Emax is the same. The increment of the natural volumetric strain of ventricle 1, $\Delta V/V_1$, is greater than that of ventricle 2, since the volume of ventricle 1 is smaller than that of ventricle 2. The pressure P is a component of stress, hence we expect that the stress-strain relation in terms of P and V should take the form $\Delta P = k_1\ \Delta V/V_1 = k_2\ \Delta V/V_2$, where k_1 and k_2 should be intrinsic elastic constants of the myocardium. Since V_1 is smaller than V_2, k_1 should be smaller than k_2. On the contrary, if both the ventricles have the same intrinsic myocardial elastic constant, that is if $k_1 = k_2$, Emax should be greater for ventricle 1, which has a smaller volume. Thus, Emax decreases with ventricular size.

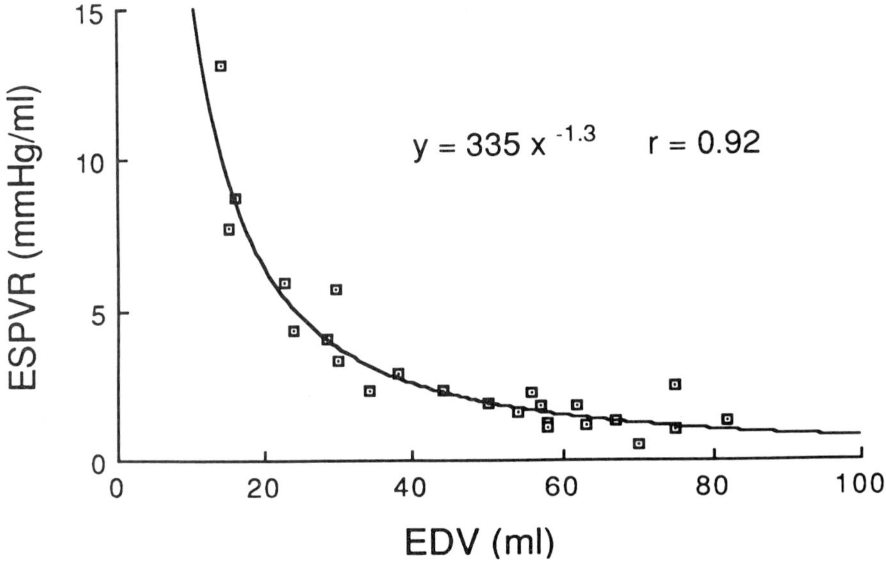

Fig. 1. Relation between Emax, the slope of end-systolic pressure-volume relation (ESPVR), and end-diastolic volume (EDV) in 25 β-blockaded dogs. Reprinted with permission from ref. 1.

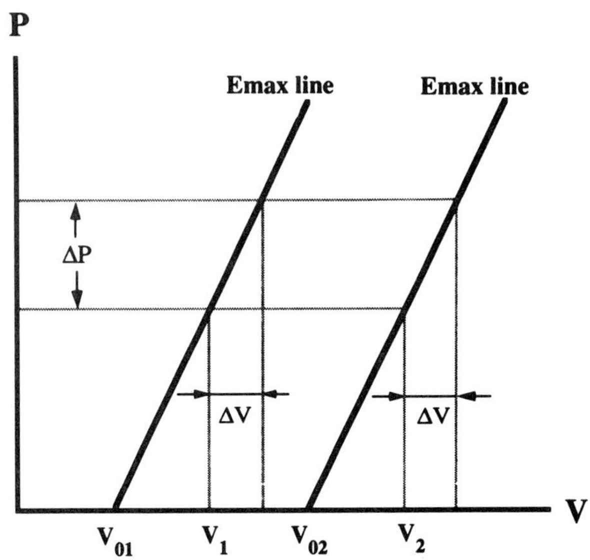

Fig. 2. End-systolic pressure-volume relations of two ventricles of differing volume but with the same Emax.

CAN EMAX BE NORMALIZED BY THE CAVITY VOLUME OR THE MASS OF THE WALL OF THE VENTRICLE?

For simplicity, we will consider a spherical model. Let us take a segment of length l on the spherical surface. The increment of linear strain of the segment is

$$\Delta l/l = 2\,\pi\,\Delta R/2\,\pi R = \Delta R/R.$$

Here, R is the radius of the sphere. The increment of areal strain of the spherical surface, that is the fractional extension of the surface, $\Delta A/A$, is given by

$$\Delta A/A = 8\,\pi R\,\Delta R/4\,\pi R^2 = 2\,\Delta R/R = 2\,\Delta l/l.$$

The increment of volumetric strain of the sphere, $\Delta V/V$, is given by

$$\Delta V/V = 4\pi R^2 \Delta R/(4\pi R^3/3) = 3\Delta R/R = 3\Delta l/l.$$

The increment of the areal strain equals two times that of the linear strain and the increment of the volumetric strain equals three times that of the linear strain. In the final form, the increment of linear strain of a segment l on the spherical surface is given as one third of the increment of volumetric strain of the sphere. Therefore, the increment of endocardial linear strain equals one third of the increment of cavity volume strain.

According to the incompressibility of the myocardium, the volume of the epicardial sphere with radius R_2 minus the volume of the endocardial sphere with radius R_1 is constant(Fig. 3), where

$$(4/3)\pi(R_2^3 - R_1^3) = \text{constant}.$$

From this relation we obtain

$$R_2^2\,\Delta R_2 - R_1^2 \Delta R_1 = 0.$$

Then, the increment of epicardial linear strain $(\Delta l/l\,)_{epi}$ is written as

$$(\Delta l/l)_{epi} = \Delta R_2/R_2 = (R_1/R_2\,)^3\,\Delta R_1\,/R_1 = (1/3)(R_1/R_2\,)^3\;\Delta V/V.$$

Similarly, the increment of midwall layer linear strain $(\Delta l/l)_{mid}$ is given by

$$(\Delta l/l)_{mid} = (1/3)[R_1/(R_1 + H/2)]^3\,\Delta V/V.$$

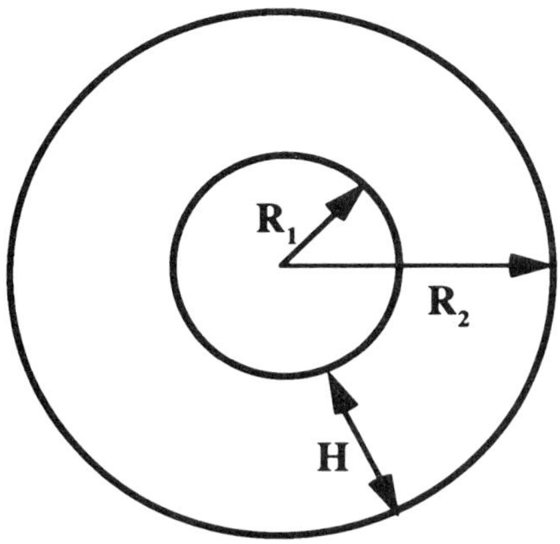

Fig. 3. A thick-walled spherical model of the ventricle with endocardial radius R_1 and epicardial radius R_2.

These strains include the epicardial radius R_2 and wall thickness H in addition to the increment of volumetric strain of the cavity.

As a stress corresponding to these strains, let us consider mean wall stress, σ which is given by

$$\sigma = PR_1^2 / 2H(R_1 + H/2).$$

Substitution of the end-systolic pressure-volume relation, $P = Emax(V - V_0)$, gives the end-systolic mean wall stress

$$\sigma = Emax(V - V_0)R_1^2 / 2H(R_1 + H/2).$$

Both the strains and the stress contain the cavity volume V and V_0, the internal radius R_1, and the wall thickness H. We can expect that there should be a type of stress-strain relation between these, that is, the increment of σ equals some constant k times the increment of linear strain. If, from this type of relation, we obtain a relation between Emax and k which includes only Vo and R_1, we can normalize Emax by the cavity volume. However, the relations we find always includes the wall thickness H. This means that Emax cannot be normalized by the cavity volume. Since the mass of the ventricular wall cannot represent the most important geometrical feature, the cavity volume, it is obvious that Emax cannot be normalized by mass.

HOW TO NORMALIZE EMAX, WORK AND PVA

The straightforward application of the theory of solid mechanics to describe the state of stress and strain in a deformable body is the expression of the components of stress and strain in terms of the stress and strain matrices. At present, however, there are no methods for measuring the stress-matrix components directly. Therefore, we must rely on mathematical models to estimate the stress-matrix components. Furthermore, there are both quantitative and qualitative discrepancies among the stress distributions calculated for the various ventricular models, and it is difficult to judge which is the most realistic representation of the actual stress distribution in the ventricular wall. Hence, such detailed analyses as stress matrix are not practically useful.

There is another reason why it is difficult to obtain a simple relation between stress and strain in a single direction in the ventricular wall. Figure 4 illustrates elongation of a muscle fiber, on each end of which a force F is applied. The longitudinal stress across any cross section of the fiber is easily obtained as F divided by the cross-sectional area of the fiber. The longitudinal strain is the ratio of extension along the fiber. The ratio of this stress to this strain is a constant of material, i.e., it is independent of the force F, the cross-sectional area and the length of the fiber. This constant is known as Young's modulus. However, things are not so simple in the ventricular wall. Parallel muscle fibers which constitute the ventricular wall are connected to each other side-by-side. When muscle fibers are elongated, their cross-sectional area decreases, and their side walls pull each other. When muscle fibers contract, their cross-sectional area increases, and their side walls push each other. Therefore, changes in longitudinal stress and strain in the ventricular wall necessarily cause changes in lateral stress and strain. Hence, it is difficult to obtain a simple relation between the components of stress and strain in one direction alone.

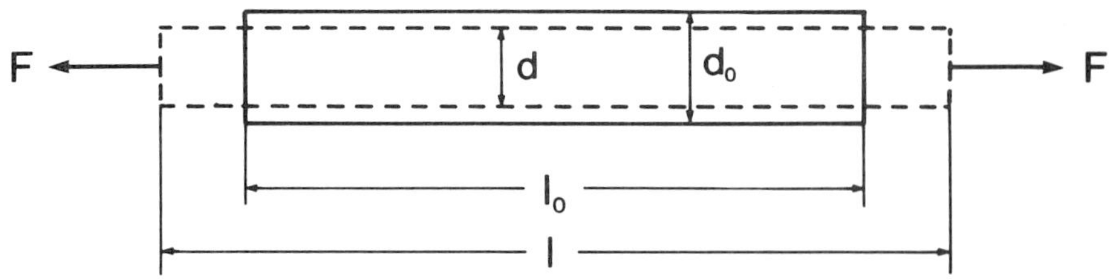

Fig. 4. Longitudinal and transverse strains caused by a longitudinal stress. F, force; d, diameter; d_0, diameter at zero force; 1,length; 1_0, length at zero force. Reprinted with permission from ref. 5.

To avoid this difficulty we have introduced a simple relation between an average stress and an average strain(2-6). The average stress and strain we used include components in two directions. Consider a region of the ventricular wall with midwall-layer area A and wall thickness H (Fig. 5). During contraction, A decreases, while H increases. During relaxation, A increases, while H decreases.

According to the incompressibility of the myocardium, the volume of the region, V_m, which is given by A times H is constant through a cardiac cycle

$$V_m = A \times H = A' \times H' = \text{constant}.$$

Therefore

$$\Delta V_m = H\Delta A + A\Delta H = 0.$$

Division of the above equation by AH gives

$$\Delta A/A + \Delta H/H = 0.$$

Using this relation we find that

$$\Delta A/A = \Delta(\ln A) = -\Delta H/H = -\Delta(\ln H) = \Delta[\ln(1/H)].$$

The above equation shows that the increment of areal strain $\Delta A/A$, which is equal to the increment of lnA, is equal to the increment of the natural logarithm of the reciprocal of wall thickness H. Therefore, we can regard ln (1/H) as areal strain.

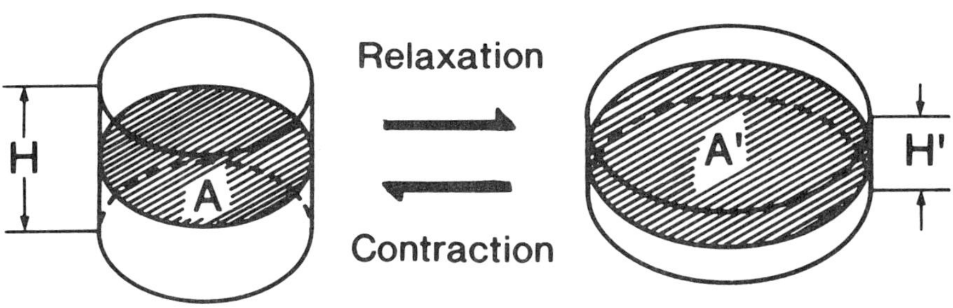

Fig. 5. Deformation of a region of the ventricular wall. A, midwall-layer area; H, wall thickness.

The following equation shows our method (2-6) of calculating regional work per unit volume of myocardium using the combination of mean wall stress σ and areal strain ln(1/H), where

$$RWM = -\int \sigma \, d[\ln(1/H)],$$

RWM is the regional stroke work per unit volume of myocardium[mJ/m^3], H is the wall thickness[cm] and σ is the mean wall stress[10^3N/m^2 = kPa]. The integral is taken over a cardiac cycle. The mean wall stress is the average of circumferential and meridional components of wall stress. In a spherical model of the ventricle, σ is given by

$$\sigma = PD/4H,$$

where P is the left ventricular pressure[10^3N/m^2 = kPa] and D is the left ventricular internal diameter[cm]. The areal strain is equal to the sum of circumferential and meridional components of wall strain. Therefore, the σ -ln(1/H) relation is not a stress-strain relation in one direction, but a relation combining two components of stress and strain.

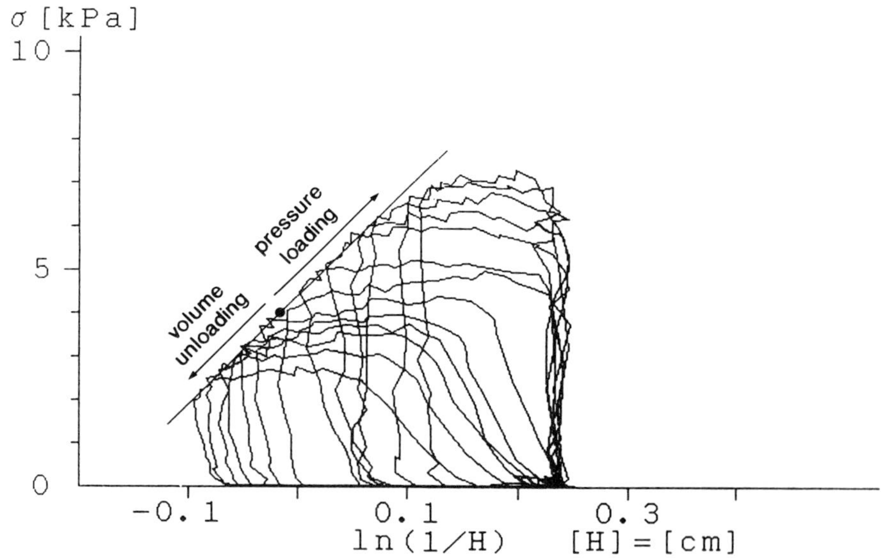

Fig. 6. Representative changes in the mean wall stress-natural logarithm of the reciprocal of wall thickness [σ-ln(1/H)] loop during volume unloading and pressure loading from a dog. σ, mean wall stress; H, wall thickness. Reprinted with permission from ref 5.

Figure 6 demonstrates the σ -ln(1/H) relation obtained from a dog during pressure loading and volume unloading. The area surrounded by a loop equals regional work per unit volume of myocardium. The end-systolic σ -ln(1/H) data points moved on the same line during both loadings. The slope of this line was increased by positive inotropic agents, and was decreased by negative inotropic agents(5,6). The physical meaning of the slope will be clarified later. Let us first validate the regional work given by this method. Figure 7 shows the simultaneous recordings of the left ventricular pressure and M-mode echocardiogram from a human, and the tracings digitized with a hand-controlled cursor. This data gives changes in the wall thickness of the interventricular septum and the left ventricular posterior wall and in the short axis diameter. Using this data we can obtain the σ -ln(1/H) relation. Figure 8 illustrates the pressure-volume curve, and the σ -ln(1/H) relation obtained from the data in Fig. 7. The σ -ln(1/H) loop for the septum is delineated by the solid line, and that for the posterior wall is the dotted loop. In this case, the posterior wall and the septum were doing nearly the same amount of regional work per unit volume of myocardium.

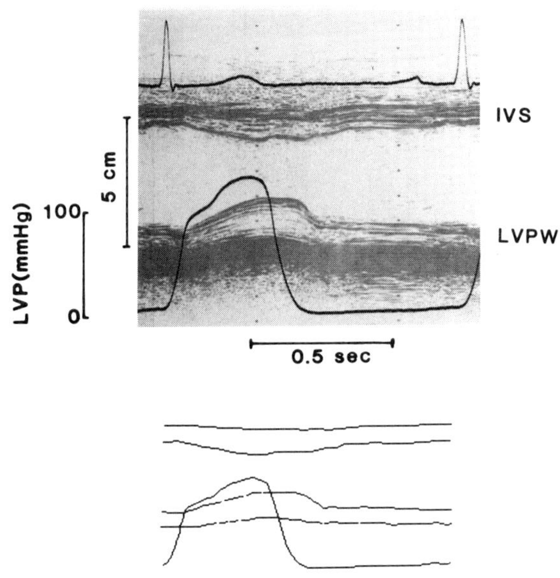

Fig. 7. Original recordings of the left ventricular pressure (LVP) and M-mode echocardiogram from a normal subject (top) and the tracings digitized with a hand-controlled cursor (bottom). IVS, interventricular septum; LVPW, left ventricular posterior wall. Reprinted with permission from ref 4.

According to the law of conservation of energy, the total regional work accumulated over the myocardial volume of the left ventricle must correspond to the total mechanical work performed by the entire left ventricle, which is obtained from the pressure volume curve. We took the average of the septal and posterior wall regional work, and multiplied it by the left ventricular myocardial volume in ten normal humans. Figure 9 shows the relation between the average regional work per unit volume, RWM, multiplied by the left ventricular myocardial volume, LVM, and the total work, TW, which was obtained from the pressure-volume curve. Although the (RWMxLVM) is linearly correlated with TW, the value of (RWMxLVM) was slightly smaller than the value of TW. We used a spherical model for simplicity. In this case, the mean wall stress is underestimated by about 27% compared with an ellipsoidal model. The dotted line shows a prediction line corresponding to the case that (RWMxLVM) is 27% smaller than TW. This falls within the 95% confidence band of the regression relation between TW and RWMxLVM, and this line coincides well with the linear regression line, which is denoted by the solid line. This supports the validity of our method of calculating the regional work.

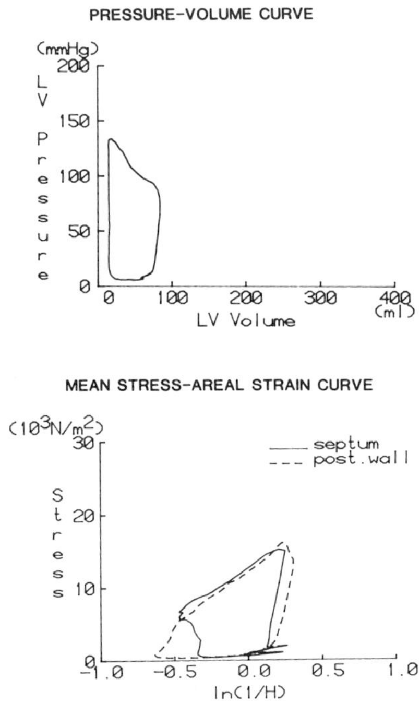

Fig. 8. The left ventricular(LV) pressure-volume curve (top) and the mean wall stress-natural logarithm of the reciprocal of wall thickness [σ-ln(1/H)] loops (bottom) for the ventricular septum (solid line) and the posterior (post.) wall of the left ventricle (dotted line) obtained from the data in Fig. 7.

The analogy of an incompressible Hookean material enables a clear understanding of the meaning of the slope of the σ -ln(1/H) relationship.

Let us consider a rectangular solid body subjected to a plane stress with components σ_x and σ_y (Fig.10). The elongations in the x- and y- directions due to σ_x and σ_y are $\epsilon_x\ \Delta x$ and $\epsilon_y\ \Delta y$, where ϵ_x and ϵ_y are the strain components. In the z-direction, contraction occurs, the amount of which is $\epsilon_z\ \Delta z$. For an incompressible Hookean material

$$\epsilon_x + \epsilon_y + \epsilon_z = 0, \text{ or } \quad \epsilon_x + \epsilon_y = -\epsilon_z$$

and the following relation between the stress components in the xy-plane and the strain component in the z-direction applies

$$(\sigma_x + \sigma_y)/2 = -E\,\epsilon_z.$$

Here, E is Young's modulus. The quantity on the left hand side of the above equation is exactly the same as our mean wall stress σ. Therefore

$$\sigma = -E\epsilon_z. \qquad (1)$$

The strain components ϵ_z is defined as ln(H/Ho), where Ho is the unstressed wall thickness. However, as far as the increment of ϵ_z is concerned, the unstressed thickness Ho can be omitted. Namely

$$d[\ln(1/H)] = -d\epsilon_z.$$

Therefore, we use ln(1/H) as a substitute for $-\epsilon_z$, which is equal to $\epsilon_x + \epsilon_y$, the areal strain. Differentiation of Eq.(1) with respect to ϵ_z yields:

$$d\sigma/d[\ln(1/H)] = E$$

The above equation shows that the slope of the σ -ln(1/H) relation gives Young's modulus E for a linearly elastic (Hookean) material(5,6).

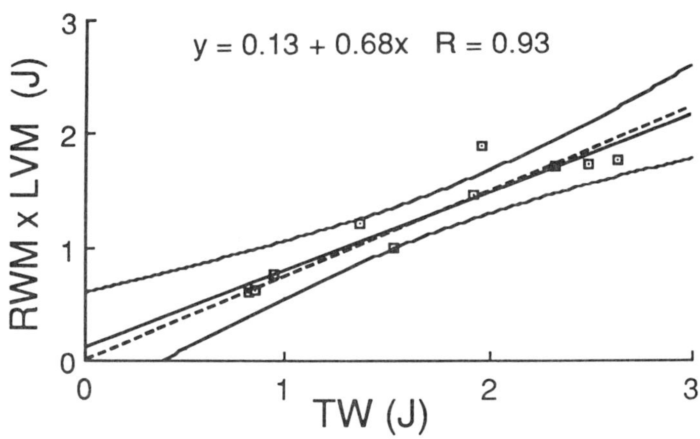

Fig. 9. Relation between the average regional work per unit volume multiplied by the left ventricular myocardial volume (RWMxLVM) and total work (TW) in the normal group (n = 10). The solid line is the linear regression line. The dotted line is the RWMxLVM versus TW relation predicted if spherical wall stress underestimates ellipsoid stress by 27%. The curved lines are the edges of the 95% confidence band for the actual RWMxLVM versus TW relation. LVM, left ventricular myocardial volume; RWM, average of regional work of interventricular septum and posterior wall of the left ventricle; TW, total work performed by the entire left ventricle. Reprinted with permission from ref. 4.

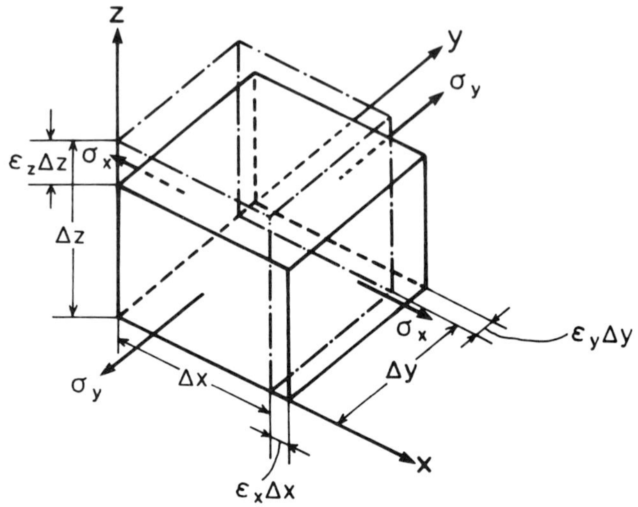

Fig. 10. Rectangular solid body subjected to a plane stress. The rectangular solid delineated by a chain dotted line shows the body before deformation, and that delineated by a solid line shows the body after deformation. σ_x,σ_y, stresses. Reprinted with permission from ref 6.

The myocardium does not necessarily exhibit a linear relation between stress σ and strain ln(1/H)(1,5). If the stress-strain relation is curvilinear, Young's modulus is replaced by the term "tangent modulus" or "elastic stiffness", which defines the slope at any point of a stress-strain curve. In such a case, the stiffness-stress relations express elastic properties of a material. Note that in the

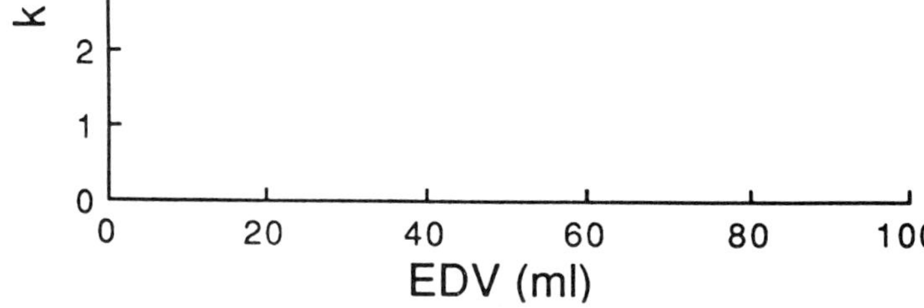

Fig. 11. Relation between the end-systolic elastic stiffness constant of the myocardium (k SM) and end-diastolic volume (EDV) in 25 β-blockaded dogs obtained from the same experiments as in Fig. 1. Reprinted with permission from ref 1.

analysis of the stiffness-stress relation and in the calculation of regional work, the unstressed wall thickness Ho is not required since ln(1/H) always appears in its differential form, d[ln(1/H)], and d[ln(Ho/H)] = d[ln(1/H)].

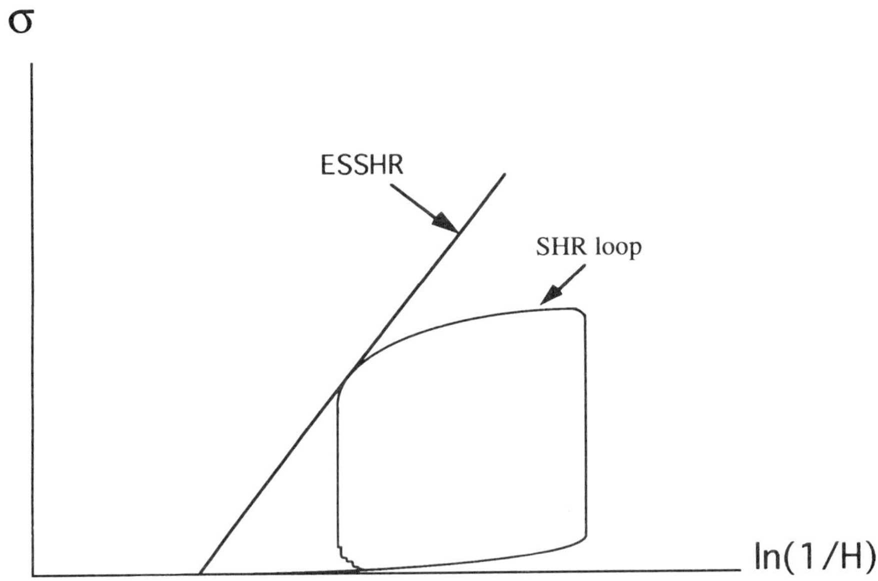

Fig. 12. Proposed pressure-volume area normalized to unit volume of myocardium based on the relation between the mean wall stress(σ) and the natural logarithm of reciprocal of wall thickness[ln(1/H)]. SHR loop, σ-ln(1/H) relation during a cardiac cycle; ESSHR, end-systolic σ-ln(1/H) relation line.

If the stress-strain relation is of the exponential form, as is often the case in biological materials

$$\sigma = C\exp[k\ln(1/H)] + B$$

and the elastic stiffness is

$$d\sigma/d[\ln(1/H)] = k\sigma - kB.$$

The elastic stiffness in this case is a linear function of the stress σ, and k is called the "stiffness constant" which represents the overall slope of the σ-ln(1/H) relation.

Figure 11 shows the relation between the stiffness constant and the end-diastolic volume obtained from the same experiments as in Fig. 1, in which we showed the dependence of Emax on ventricular size. Despite the wide range of ventricular volume, 14 ml to 82 ml, the stiffness constant was nearly constant with a narrow distribution and was independent of left ventricular size.

CONCLUSION

In conclusion, the area surrounded by the σ -ln(1/H) loop during a cardiac cycle is equal to the stroke work per unit volume of myocardium. The physical

meaning of the slope of the σ -ln(1/H) relation is the same as Young's modulus, and the end-systolic σ -ln(1/H) relation can be regarded as the normalized Emax. Figure 12 is the proposed PVA normalized to unit volume of myocardium, which is based on the relation between the mean wall stress σ and the areal strain of the midwall layer, ln(1/H).

REFERENCES

1. Nakano, K. et al. Myocardial stiffness derived from end-systolic wall stress and logarithm of reciprocal of wall thickness: Contractility index independent of ventricular size. Circulation 82:1352-1361, 1990.

2. Sugawara, M. et al. Regional work of the ventricle: Wall tension-area relation. Heart Vessels 1:133-144, 1985.

3. Nakano, K. et al. A new approach to defining regional work of the ventricle and evaluating regional cardiac function: Mean wall stress-natural logarithm of reciprocal of wall thickness relationship. Heart Vessels 2:74-80, 1986.

4. Nakano, K. et al. Regional work of the human left ventricle calculated by wall stress and the natural logarithm of reciprocal of wall thickness. J Am Coll Cardiol 12:1442-1448, 1988.

5. Sugawara, M. et al. Regional work of the left ventricle and contractility index independent of ventricular size. In:Sasayama S, Suga H ed. Recent progress in failing heart syndrome. Springer-Verlag pp 257-279, 1991.

6. Sugawara, M., Nakano, K. A new method of analyzing regional myocardial function of the ventricle. In: Hori M et al. ed. Cardiac mechanics and function in the normal and diseased heart. Springer-Verlag pp 249-256, 1989.

7. Observations on the Relation of PVA and MVO_2 in Closed Chest Dogs

Gregory L. Freeman and Sumanth D. Prabhu

INTRODUCTION

Energy utilization is a key process underlying all cellular function. In excitable contractile tissue, such as the myocardium, energy use is particularly important since the mechanical work performed requires a high rate of energy consumption. In addition to routine needs for basal metabolism and maintenance of transmembrane gradients, cardiac tissue requires energy to support beat-to-beat cyclical calcium movement, and to allow contractile elements to generate force and to shorten. In the absence of sufficient energy, cardiac performance deteriorates rapidly.

Over the past decade the study of cardiac energetics has been significantly advanced as a result of the analytical approach developed by Suga and colleagues (See ref. 1 for review). By evaluating the mechanical work of the heart in the pressure-volume plane and relating it to oxygen consumption (MVO_2), which provides essentially all of the energy used by the heart, this framework has allowed evaluation of the energy usage of the intact ventricle. This behavior has been studied under baseline conditions as well as in response to a variety of positive and negative stimuli. The central tenets of this construct are that the total mechanical output of the ventricle, quantified as the pressure-volume area (PVA), is linearly related to MVO_2, that the slope of the MVO_2-PVA relation is inversely related to contractile efficiency, and that the Y axis intercept of the relation quantifies energy used for nonmechanical processes.

The purpose of our studies (2) has been to evaluate the MVO_2-PVA relation in chronically instrumented dogs. Use of this preparation offers the advantage of allowing longitudinal assessment of both myocardial performance and efficiency in individual animals before, during, and after development of a pathological condition. As a first step, preliminary studies were performed to determine whether energetic studies were feasible in this model, and to see whether responses of the MVO_2-PVA relation to inotropic and chronotropic stimuli are the same as those seen in the isolated heart. Our results demonstrate that the relation can be defined in this preparation, and that the response to catecholamine infusion is similar to that reported from isolated hearts; on the other hand, responses to increased heart rate differ from those seen in previous studies.

METHODS

Ten mongrel dogs were used for these studies. Details of the preparation have been previously published (3,4), and are shown in Figure 1. In brief, the animals were operated under sterile surgical conditions for implantation of instruments for chronic hemodynamic monitoring. Three sets of orthogonally oriented endocardial piezoelectric diameter gauges and a high fidelity micro-manometer (Konigsberg Instruments, Pasadena, Calif) were implanted into the left ventricle (LV). Doppler coronary flow probes were placed on the left circumflex artery in 8 dogs and both the left circumflex and the left anterior descending coronary artery in 2 dogs. Left atrial pacing wires and vena caval balloon occluder cuffs were also implanted, as were fluid filled left atrial and LV cannulae. The animals were allowed to recover from surgery for at least two weeks before they were studied.

On the day of the study the animals were taken to a radiological suite and anesthetized with a combination of thiopental sodium (25-30 mg/kg), droperidol (1.5-3.0 mg/kg), and fentanyl (0.03-0.06 mg/kg). They were intubated and ventilated with 100% O_2. A jugular cutdown was performed and a fiberoptic oximetric catheter (Opticath, Abbott, North Chicago, Illinois) was passed into the coronary sinus (CS) under fluoroscopic guidance. The animals was given heparin (5,000 U), and the catheters were regularly flushed. Autonomic blockade was produced with atropine (0.1 mg/kg) and hexamethonium (20-25 mg/kg).

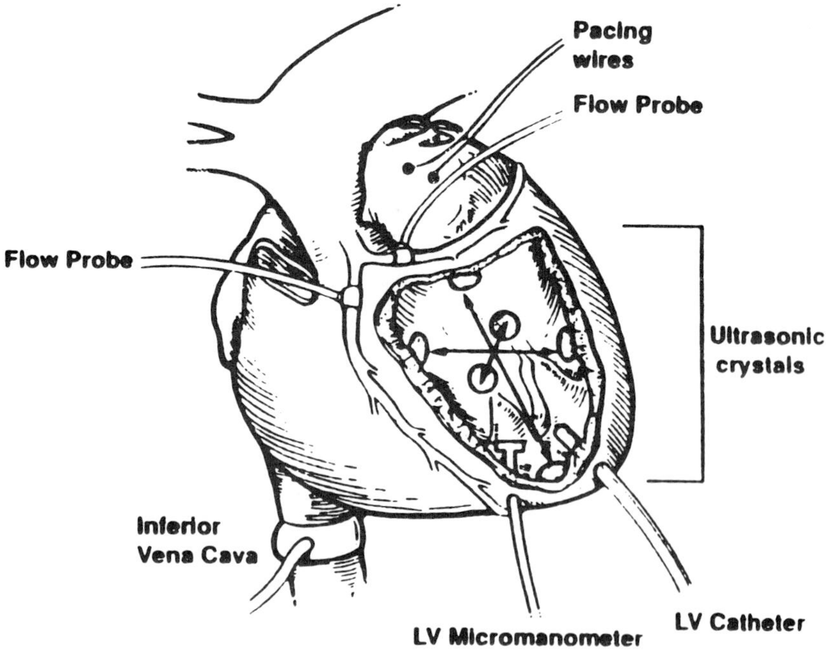

Fig. 1. Surgical instrumentation used for the experiments.

Analog tracings were made of LV pressure, the first derivative of LV pressure (dP/dt), the limb lead ECG, the three diameters, and coronary flow. These data were simultaneously converted to digital form at a rate of 500 Hz and stored in an IBM PC. CS oxygen saturation was manually noted at 5 second intervals throughout each data recording periods. Under each set of test conditions data were collected during steady-state and caval occlusion conditions. In eight dogs data were acquired at heart rates of ~120 bpm, 160 bpm and 200 bpm; three of these animals also were tested before and after 9 μg/kg/min I.V. dobutamine. Two additional dogs received the dobutamine but not the heart rate study.

At the end of the protocol the animals were killed by lethal KCl injection following pentobarbital anesthesia. The heart was dissected and indocyanine green dye was used to determine the portions of the ventricle perfused by the LCX and the LAD (unstained myocardium), which were separately weighed.

The digitized data were analyzed using programs developed in our lab. LV volume was calculated from the three diameters using the formula

$$V_{LV}=(\pi/6)D_{AP}\bullet D_{SL}\bullet D_{LA}$$

where D_{AP} is the anterior-posterior diameter, D_{SL} is the septal-lateral diameter, and D_{LA} is the long axis diameter. The end-systolic pressure volume (P_{es}-V_{es}) relation was derived using an iterative approach (5), fitting the data to the equation

$$P_{es}=E_{es}(V_{es}-V_0)$$

where E_{es} is the slope of the relation and V_0 is its pressure volume intercept. The PVA was defined as the area bounded by the pressure-volume loop and the P_{es}-V_{es} relation.

AVO_2 difference was quantified as the aortic saturation content minus the coronary sinus saturation content using the formula

$$AVO_2 \text{ difference} = 1.36 \bullet Hb \bullet (Sa_{O2}-Sc_{O2}) \bullet 10^{-2}.$$

MVO_2 was calculated using the Fick principle

$$MVO_2=CF_{avg} \bullet AVO_2 \text{ difference} \bullet HR^{-1}$$

where CF_{avg} is the average coronary flow and HR is heart rate.

Beats from the beginning, middle and end of each caval occlusion run during basal and test conditions were used to define the MVO_2-PVA relation. The relation was described by its slope (α) and MVO_2 intercept (β). To allow comparison between different animals, the values for MVO_2 and PVA were normalized per 100g LV and converted to joules (J). For animals with a single coronary flow probe, MVO_2 was normalized to the weight of myocardium in the LCX distribution, as determined at necropsy. Contractile efficiency was determined as the inverse of the slope of the MVO_2-PVA relation, and expressed as a percent.

Correlations between MVO_2 and PVA were made using standard linear regression techniques. Comparisons of hemodynamic and energetic variables between animals at different heart rates were made using repeated measures analysis of variance and the paired t test with Bonferroni correction. Comparisons before and after dobutamine were made using the paired t test.

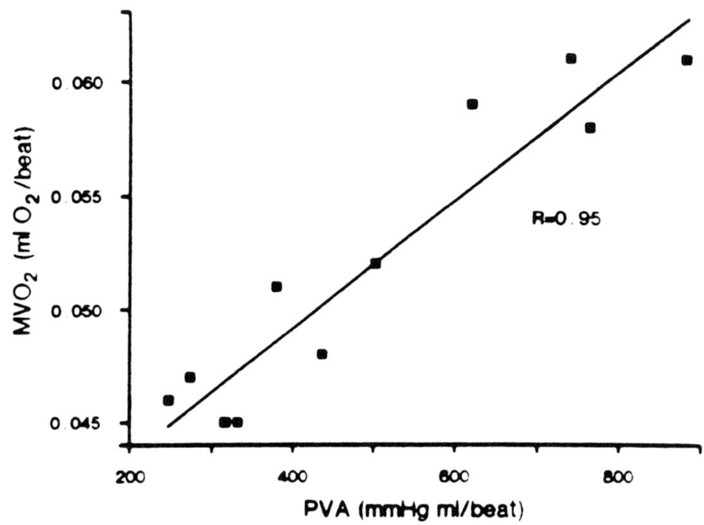

Fig. 2. Myocardial O_2 consumption (MVO_2) plotted as a function of pressure-volume area (PVA) under baseline conditions from one animal. MVO_2 is linearly related to PVA with a correlation coefficient of 0.95.

RESULTS

Figure 2 shows the MVO_2-PVA relation during control conditions from one dog. As has been shown repeatedly in isolated heart studies, MVO_2 decreased proportionally to PVA; the relation was linear with high correlation coefficients in all of the animals. For these dogs under baseline conditions, contractile efficiency was 25.6 $\pm$ 2.8 %, ranging from 16.9 to 38.4%.

Heart rate had an effect on energetics not seen in previously published studies performed on isolated hearts. As can be seen in Figure 3, at the high heart rate contractility, indexed as E_{es}, increased substantially (from 9.4 $\pm$ 1.3 to 18.6 $\pm$ 3.1 mmHg/ml, $p < 0.0167$). At this heart rate the efficiency decreased from 25.7 $\pm$ 0.32 to 16.3 $\pm$ 2.4% ($p < 0.0167$). In addition, the intercept of the relation decreased from 0.562 $\pm$ 0.119 to 0.377 $\pm$ 0.074 J/beat/100g LV, $p < 0.0167$.

Thus, we found increased contractility and decreased efficiency in response to elevated heart rate in our preparation.

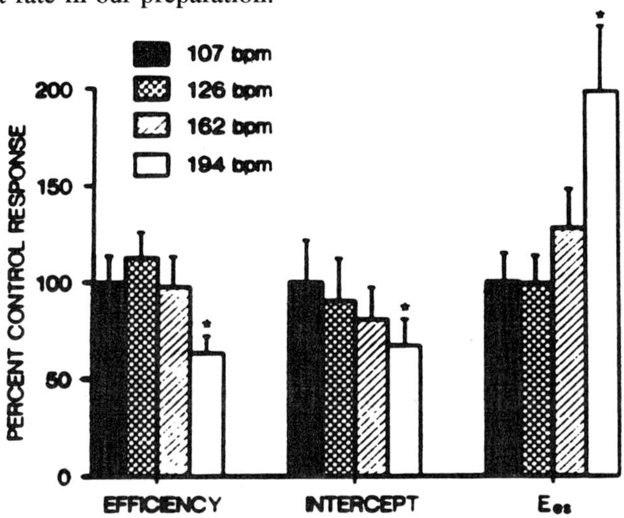

Fig. 3. Group data (n=8) for contractile efficiency, MVO_2 axis intercept (J/beat/100g LV) and elastance (E_{es}) at baseline (107±3 beats/min), low paced (126±2 beats/min), intermediate paced (162±2 beats/min), and high paced (194±5 beats/min) heart rates. Each variable is normalized to control value and expressed as a percent. High hearts rates produce significant reductions in contractile efficiency and MVO_2 axis intercept (load-independent MVO_2 per beat) and significant increases in elastance. *P < 0.0167 compared with baseline and low paced heart rates. All values are means ± SE.

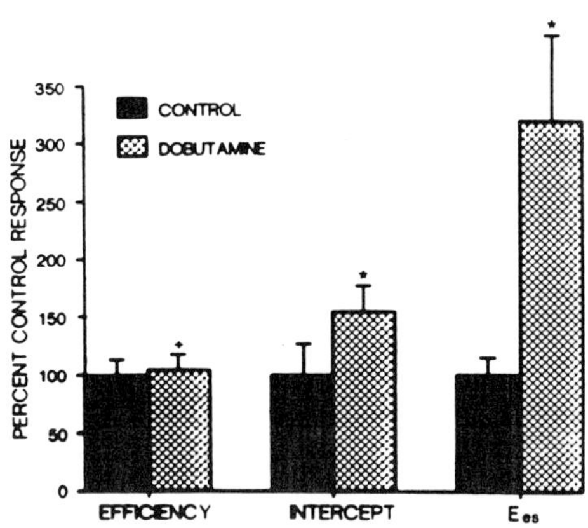

Fig. 4. Group data (n=5) for contractile efficiency, MVO_2 axis intercept (J/beat/100g LV), and E_{es} at baseline and after dobutamine administration. Each variable is normalized to control value and expressed as a percent. There is no significant change in contractile efficiency. MVO_2 axis intercept and E_{es} are significantly increased. + P = not significant compared with control. * P < 0.05 compared with control. All values are means ± SE.

To avoid confounding effects of changes in heart rate, the response of the MVO_2-PVA relation to dobutamine was evaluated during atrial pacing at 155 bpm. In three dogs the heart rate during dobutamine was slightly faster than this; the mean value during the dobutamine experiments was 167 ± 6 bpm. As can be seen in Figure 4, dobutamine had the expected effect on contractility (E_{es} increased from 10.1 ± 1.5 to 32.0 ± 7.6 mmHg/ml, $p < 0.05$). Efficiency, on the other hand, was not changed (28.8 ± 3.8 vs 30.3 ± 3.8%). In agreement with prior studies in the isolated heart, we found an increase in the intercept value from 0.392 ± 0.108 to 0.607 ± 0.083 J/beat/100 g LV ($p<0.05$).

DISCUSSION

The results of these studies show for the first time that the MVO_2-PVA relation can be accurately quantified in the chronically instrumented dog. We found that the efficiency of the LV under baseline conditions was lower than that reported for the isolated heart. In addition, while our results on the response of this relation to dobutamine confirmed observations made in isolated hearts (6), we found that heart rate had a unique effect, such that at high rates there was reduced efficiency.

It is of interest that our results regarding heart rate differ from those of studies which have evaluated isolated hearts. Suga and colleagues (7) showed that increasing rate from 124 ± 17 to 193 ± 23 bpm had no effect on contractile efficiency. In that study the slope of the MVO_2-PVA relation was $1.74 \pm 0.44 \times 10^{-5}$ ml O_2/mmHg/ml at the control heart rate, and $1.80 \pm 0.67 \times 10^{-5}$ ml O_2/mmHg/ml at rapid rates. Of note, our results also differ with respect to the effect of increased heart rate on contractile performance. In their study the slope of the P_{es}-V_{es} relation (E_{max}) was 6.4 ± 1.9 mmHg/ml/100 g LV at control rates, and 7.0 ± 1.7 mmHg/ml/100 g LV at rapid rates. This is in distinct contrast to our findings in the present study, where E_{es} increased from 9.4 ± 1.3 at control rates to 18.6 ± 3.1 mmHg/ml at rapid rates. Absence of increased contractility in response to heart rate stimulus had previously been shown in the isolated heart by this group (8).

Our results also differ from those of Suga and colleagues with regard to the impact of heart rate on oxygen used for nonmechanical needs (basal metabolism and excitation-contraction coupling). We found a reduction in the nonmechanical MVO_2 used per beat, which contrasts with data reported from isolated hearts (7), where no change in unloaded MVO_2 per beat was found. Since basal metabolic rate per minute is constant (9), the requirements for basal metabolism will be less for a given beat at the higher heart rate (each beat occurs in less time). To account for the fact that at high heart rates nonmechanical MVO_2 was unchanged despite reduced per beat expenditure for basal metabolism, Suga et al proposed that there must be a proportional increase in energy requirements for excitation-contraction coupling per beat (7). This is consistent with both an increased number of beats per minute, and

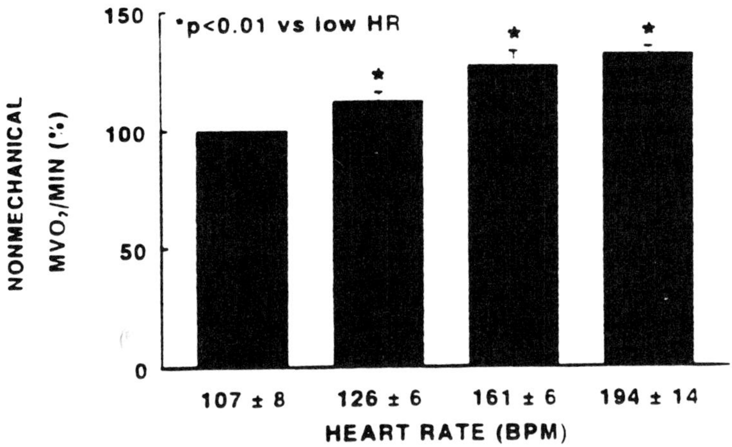

Fig. 5. MVO_2 used for nonmechanical processes per minute, expressed as a percentage of the value at the lowest heart rate.

with the known increase in intracellular calcium transients found at high heart rates (10). Of interest, we also found that overall nonmechanical MVO_2 used per minute went up at high heart rates (Figure 5), which is consistent with increased energy use for excitation-contraction coupling. However, the magnitude of the increase was less in our preparation, since the nonmechanical MVO_2 per beat decreased at high heart rates.

In two previous studies we have demonstrated that the force-frequency effect clearly plays an important role in modulating contractile performance in the intact animal. In the first (11), we showed that the slope of the P_{es}-V_{es} relation was tightly correlated to heart rate, with a Pearson correlation coefficient of 0.96. E_{max} increased to 238 ± 99% of control values at peak rates of 180 to 200 bpm. Likewise, dP/dt_{max} at a matched LV end-diastolic volume increased 487 ± 50 mmHg/sec/100 bpm, confirming the importance of heart rate on contractility in the intact animal. Similar results were found in two animals subjected to autonomic blockade, confirming that this property is intrinsic to the myocardium.

In the second (12) study we confirmed the influence of heart rate on contractility. In addition it was shown that there was a cumulative effect of heart rate and dobutamine on contractility, suggesting that these forms of inotropic stimulation may have unique, and possibly additive mechanisms. Of particular note, heart rate also had a characteristically different influence on ventriculovascular coupling, such that there was a substantially greater increase in effective arterial elastance with pacing than after dobutamine. This was quantified by the percentage of PVA expressed as stroke work, which was markedly larger after dobutamine infusion than during pacing. Clearly, the results of these two studies demonstrate that the force-frequency response is active in the intact animal.

Thus, while our results on the way in which the MVO_2-PVA relation responds to increased heart rate differ from those of investigators using isolated hearts, our results on basic mechanical response to increased heart rate do as well, suggesting this observation represents a true physiological phenomenon. Moreover, our results on the effect of catecholamine infusion (dobutamine) on the MVO_2-PVA relation correspond to those found in isolated hearts, where similar inotropic results are found. The question regarding why we see differences with respect to chronotropic stimuli remains unanswered, and is of interest since the long term goal of studies of ventricular performance is to better understand function in the physiological setting.

A well known case in which the mechanical impact of increased heart rate is blunted is heart failure, where the force-frequency response has been shown to be impaired. In samples of myocardium taken from failing human hearts, Mulieri and colleagues (13) showed that contractile response did not change over a wide range of stimulation frequencies (60 to 150 per minute), whereas in normal myocardium there was a 107% increase in force production. Ezzaher et al. (14) showed that in rabbits with heart failure the normal positive staircase response to increased stimulation frequency was replaced by a negative staircase. Absence of a positive force-frequency effect in isolated, supported hearts (7,8) could result from a degree of baseline contractile depression inherent in the preparation, which may effect the myocardial response to this form of inotropic stimuli more than its response to catecholamines. Of note, other quantitative differences are found when studying isolated and intact hearts. In prior studies of the force-interval response, another frequency-dependent phenomenon, we found that the time constant of mechanical restitution was much shorter in closed-chest dogs (15) than had previously been reported in isolated hearts (16). Thus, divergences in our results regarding energetic aspects of the force-frequency response may derive from differences in the preparations; further studies will be needed to fully resolve these questions.

Our confirmation of a parallel upward shift of the MVO_2-PVA relation in response to dobutamine supports the notion that this form of inotropic stimulation does not influence mechanoenergetic performance, and serves to largely increase the energy used for nonmechanical requirements. We cannot specifically address whether this results from increased basal metabolic rate or a higher energy cost of excitation-contraction coupling.

The most critical technical question regarding our preparation is our use of data taken during acutely altered loading conditions, as opposed to steady-state conditions. We have chosen to use this approach because we feel it is most accurate for assessing the mechanical state of the myocardium at a particular point in time. We (17,18) and others (19) have shown that the P_{es}-V_{es} relation is shifted by sustained changes in afterload. This would clearly impact on PVA measured in our preparation. Indeed, we have shown that the steady-state composite P_{es}-V_{es} relation differs substantially from the acutely generated relation.

In a recent paper Nozawa, et al (20) confirmed that these relations can be quantified in the chronically instrumented dog. In that study steady-state data were compared to transient data recorded under the assumption that oxygen content of the coronary blood was unchanged during a caval occlusion (this is a reasonable assumption: in our studies the coronary sinus O_2 saturation changed only a few percent over the course of a vena caval occlusion). While the authors state that the techniques yield different results, this conclusion is difficult to apply to our study, since their assessment of energetics during caval occlusion lumped data from three disparate sets of loading conditions (See Figure 4 of ref 18). A true comparison would require using only baseline caval occlusion data, which appear from their Figure 4, to lie parallel to the steady-state line. Further assessment of the impact of this analytical approach will be needed.

In conclusion, our studies have demonstrated the usefulness of the chronically instrumented dog preparation for evaluating the mechanoenergetic function of the left ventricle. Our results indicate that baseline contractile efficiency is lower in the intact animal than generally described in the isolated heart. Tachycardia reduces contractile efficiency and unloaded MVO_2 per beat. Dobutamine increased the MVO_2 intercept while maintaining efficiency. Both interventions increase the per minute O_2 requirements for nonmechanical processes, and this is of much greater magnitude with dobutamine. While our observations confirm some aspects of the prior work from isolated hearts, it opens questions regarding the impact of increased heart rate on energy efficiency in the intact heart. Future studies will allow longitudinal assessment of animals subjected to a variety of pathophysiological perturbations. This will more fully expand the important analytical framework of the pressure-volume plane introduced by Dr. Suga.

ACKNOWLEDGEMENTS

This work was supported by the research service of the Department of Veterans Affairs and by a Grant-in-Aid from the Texas Affiliate of the American Heart Association. We thank Danny Escobedo and Cindy Garcia for technical assistance, Abbott Laboratories for providing the oximetric catheter computer, Merardo Monzon for software development, and John Schoolfield for statistical advice.

REFERENCES

1. Suga, H. Ventricular energetics. Physiol Rev 79:247-277, 1990.

2. Prabhu, S.D., Freeman, G.L. Left ventricular energetics in closed-chest dogs. Am J Physiol 265:H1048-H1055, 1993.

3. Sodums, M.T., Badke, F.R., Starling, M.R., Little, W.C., O'Rourke, R.A. Evaluation of left ventricular contractile performance utilizing end-systolic pressure-volume relationships in conscious dogs. Circ Res 54:731-739, 1984.

4. Freeman, G.L., Watkins, D., Little, W.C. A simple method for measurement of left ventricular septal-lateral dimension. Am J Physiol 254:H801-H803, 1988.

5. Kono, A., Maughan, W.L., Sunagawa, K., Hamilton, K., Sagawa, K.,Weisfeldt, M.L. The use of left ventricular end-ejection pressure and peak pressure in the estimation of the end-systolic pressure-volume relationship. Circulation 70:1057-1065, 1984.

6. Suga, H., Hisano, R., Goto, Y., Yamada, O., Igarashi, Y. Effect of positive inotropic agents on the relation between oxygen consumption and systolic pressure-volume area in canine left ventricle. Circ Res 53:306-318, 1983.

7. Suga, H., Hisano, R., Hirata, S., Hayashi, T., Yamada, O., Ninomiya, I. Heart rate-independent energetics and systolic pressure-volume area in dog heart. Am J Physiol 244:H206-H214, 1983.

8. Suga, H., Sagawa, K., Shoukas, A.A. Load independence of the instantaneous pressure-volume ratio of the canine left ventricle and effects of epinephrine and heart rate on the ratio. Circ Res 32:314-322, 1973.

9. Gibbs, C.L., Chapman, J.B. Cardiac energetics. In: *Handbook of Physiology. The Cardiovascular System. the Heart*. Bethesda, MD: Am Physiol Soc, vol I: 775-804, 1979.

10. Morgan, J.P., Blinks, J.R. Intracellular Ca^{2+} transients in the cat papillary muscle. Can J Physiol Pharmacol 60:524-528, 1982.

11. Freeman, G.L., Little, W.C., O'Rourke, R.A. Influence of heart rate on left ventricular performance in conscious dogs. Circ Res 61:455-464, 1987.

12. Freeman, G.L., Colston, J.T. Role of ventriculovascular coupling in cardiac response to increased contractility in closed-chest dog. J Clin Invest 86:1278-1284,1990.

13. Mulieri, L.A., Hasenfuss, G., Leavitt, B., Allen, P.D., Alpert, N.R. Altered myocardial force-frequency relation in human heart failure. Circulation 85:1743-1750, 1992.

14. Ezzaher, A., el Houda Bouanani, N., Crozatier, B. Force-frequency relations and response to ryanodine in failing rabbit hearts. Am J Physiol 263:H1710-H1715, 1992.

15. Freeman, G.L., Colston, J.T. Evaluation of left ventricular mechanical restitution in closed-chest dogs based on single-beat elastance. Circ Res 67: 1437-1445, 1990.

16. Burkhoff, D., Yue, D.T., Franz, M.R., Hunter, W.C., Sagawa, K. Mechanical restitution of isolated perfused canine left ventricles. Am J Physiol 246:H8-H16, 1984.

17. Freeman, G.L., Little, W.C., O'Rourke, R.A. The effect of vasoactive agents on the left ventricular end-systolic pressure-volume relation in closed-chest dogs. Circulation 74:1107-1113, 1986.

18. Freeman, G.L. Effects of increased afterload on left ventricular function in closed-chest dogs. Am J Physiol 259:H619-H625, 1990.

19. Baan, J., van der Velde, E.T. Sensitivity of left ventricular end-systolic pressure-volume relation to type of loading intervention in dogs. Circ Res 62: 1247-1258, 1988.

20. Nozawa, T., Cheng, C-P., Noda, T., Little, W.C. Relation between left ventricular oxygen consumption and pressure-volume area in conscious dogs. Circulation 89:810-817, 1994.

8. Regional Myocardial Contraction Coupled With Energetics

Yoichi Goto, Satoshi Yasuda, Toshiaki Shishido, Kengo Fukushima and Shogo Suzuki

INTRODUCTION

Assessment of Regional Myocardial Function

Diagnosis and treatment of patients with coronary artery disease should be based on correct assessment of left ventricular regional contractile function because of the nonuniform nature of contractile dysfunction in these patients. Also, patients with hypertrophic (1) or dilated cardiomyopathy (2) have been reported to demonstrate regional abnormalities in contraction. Global indexes of left ventricular function (for example, ejection fraction) cannot detect such regional abnormalities of myocardial contractile function (3). To assess regional myocardial contractile function, many indexes have been proposed, as listed in Table 1.

Table 1. Indexes of regional contractile function.

1) Indexes of extent of myocardial shortening
   segment shortening, chord shortening, fractional area shortening (area shrinkage), wall thickening
2) Indexes of regional work
   pressure-length loop, pressure-thickness loop, wall tension-area loop, wall stress-1n(1/thickness) loop, preload-recruitable stroke work index
3) Indexes at end-systole
   end-systolic pressure-length relation, end-systolic pressure-thickness relation, end-systolic tension-area relation, end-systolic stress-1n(1/thickness) relation

Perfusion-Contraction Matching

Among these indexes, the amount of myocardial shortening (1,2), wall thickening (3), and the regional work index (4) have been conventionally utilized to quantify regional contractile function. All these variables decrease during myocardial ischemia (2-4), and Ross (5) has recently suggested that there is a tight coupling between oxygen supply and regional myocardial contraction ("myocardial perfusion-contraction matching"). However, the relation between regional myocardial oxygen consumption (VO_2) and variables of regional myocardial contractile function under various loading conditions or with differences in regional myocardial contractility is not fully understood.

Ventricular Energetics

In contrast, the relation between left ventricular global VO_2 and contractile function has been extensively studied using the concept of E_{max} (a left ventricular contractility

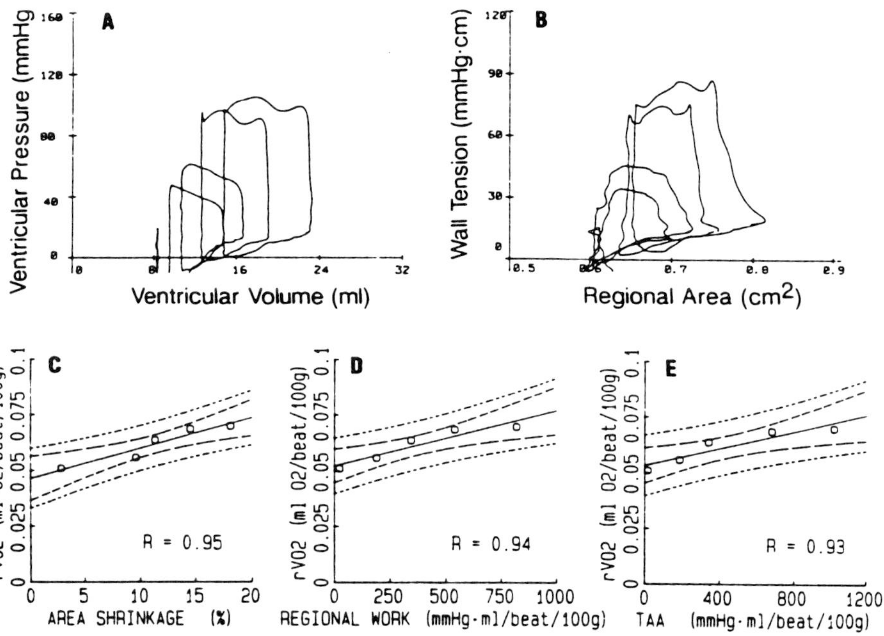

Fig. 1. Methods of analyzing left ventricular global (A) and regional (B) mechanoenergetics using the pressure-volume (P-V) and the wall tension-regional area (T-A) diagrams. External work performed by the whole ventricle and by a region can be assessed from the areas within the P-V and T-A loops, respectively. The total mechanical energy is represented by systolic pressure-volume area (PVA) for the whole ventricle and tension-area area (TAA) for the region. Both PVA and TAA consist of two portions, external work (EW) and potential energy (PE), and have the dimensions of energy because 1 mmHg·cm^3 = 1.22 x 10^{-4} Joule. ESPVR, end-systolic P-V relation; EDPVR, end-diastolic P-V relation; ESTAR, end-systolic T-A relation; EDTAR, end-diastolic T-A relation. Reprinted with permission from ref. 14.

index) and systolic pressure-volume area (PVA) (6) in the pressure-volume diagram (Figure 1A). Previous studies from our (6,7) and other laboratories (8,9) indicate that PVA linearly correlates with global VO_2 per beat under various loading conditions. PVA is the area circumscribed by the end-systolic and end-diastolic pressure-volume relations and the systolic pressure-volume trajectory. It consists of two components, external work and end-systolic potential energy (Figure 1A), and theoretically represents total mechanical energy generated by a ventricular contraction (6-8).

Regional Energetics

Recently, we have analyzed left ventricular regional contractile function using the wall tension-regional area (T-A) loop method (10-12) and demonstrated that left ventricular regional work can be reliably quantified by the area within a left ventricular wall tension-regional area (T-A) loop (Figure 1B). Because this T-A loop area has the dimensions of energy (joules), we consider that the T-A diagram for a ventricular region is energetically equivalent to the pressure-volume diagram for the whole ventricle (13). Therefore, by analogy to left ventricular PVA, systolic "wall tension-regional area" area (TAA) can be defined in the T-A diagram as a measure of regional total mechanical energy generated by a regional myocardial contraction (Figure 1B). The next step is to examine the relation between TAA and regional energy metabolism. In this article, we present our data on the relation between TAA and regional VO_2 under varying loading conditions and contractility to determine the coupling between regional myocardial contraction and energetics (14).

METHODS

Heart Preparation

Experiments were performed on the left ventricles of isolated, cross-circulated (blood-perfused) dog hearts connected to a volume servo pump system (7,10,11). A proximal portion of the left anterior descending coronary artery was cannulated and coronary blood flow was measured with an electromagnetic flowmeter. The proximal portion of the anterior interventricular vein was also cannulated (15) and coronary arteriovenous oxygen content difference between the anterior descending artery and the anterior interventricular vein was measured continuously. Orthogonal subendocardial segment lengths (i.e., circumferential and longitudinal segments) within the perfusion area of the cannulated left anterior descending coronary artery were measured with two pairs of ultrasonic crystals (11,12).

Experimental Protocol

Loading Condition Protocol

The relation between left ventricular regional VO_2 and regional myocardial contractile function (i.e., regional area shrinkage, regional work, and TAA) was assessed under four loading condition runs generated with the volume servo pump, as follows: 1) Increased preload run: increased end-diastolic and stroke volumes with a constant ejection fraction; 2) Decreased afterload run: increased stroke volume at a constant end-diastolic volume; 3) Constant stroke volume run: increased end-diastolic and end-systolic volumes with a constant stroke volume. 4) Contraction mode run: Ejecting vs isovolumic contractions.

Regional Contractility Protocol

The effect of enhancement of regional myocardial contractility on the relation between regional VO_2 and TAA was assessed. Regional myocardial contractility was selectively enhanced in the anterior wall by infusing dobutamine into the left anterior descending coronary artery.

Data Analysis

Wall tension (force per unit length, T, in $mmHg \cdot cm^{-1}$) was calculated according to the force equilibrium relation of a thick walled sphere (10-12,16) and regional area (cm^2) was calculated as the product of the two orthogonal segment lengths. T-A area (TAA, in $mmHg \cdot ml \cdot beat^{-1} \cdot 100g^{-1}$) was defined in the T-A diagram as a specific area surrounded by the end-systolic T-A relation (ESTAR), the end-diastolic T-A relation and the systolic T-A trajectory (Figure 1B). Regional work ($mmHg \cdot ml \cdot beat^{-1} \cdot 100g^{-1}$) was assessed from the integral of wall tension with respect to the area of each region, i.e., area within a T-A loop (10-12). Regional area shrinkage (%), an index of myocardial shortening, was calculated as coronary blood flow times arteriovenous oxygen content difference (vol%) between the left anterior descending coronary artery and the anterior interventricular vein (15) divided by heart rate, and expressed as ml $O_2 \cdot beat^{-1} \cdot 100g^{-1}$.

RESULTS

Loading Condition Protocol

Figure 2 (A and B) shows representative examples of pressure-volume loops (A) and T-A loops (B) simultaneously measured during the increased preload run. Both end-diastolic and stroke volumes were gradually increased while keeping ventricular ejection fraction nearly constant at 33-35%. Under these loading conditions, regional area shrinkage, regional work, and TAA positively correlated with regional VO_2 (Figure 2, C-E). Figure 3 shows representative examples of pressure-volume loops (A) and T-A loops (B) obtained during the decreased afterload run. When stroke volume was gradually increased starting from an isovolumic contraction while keeping end-diastolic volume nearly constant, systolic pressure and wall tension progressively decreased. Under these reduced afterload conditions, regional area shrinkage showed a negative correlation with regional VO_2 (Figure 3C), while regional work did not correlate significantly with regional VO_2 (Figure 3D). In contrast, TAA showed a highly linear, positive correlation despite the great changes in both ventricular and regional afterloads (Figure 3E). In the constant stroke

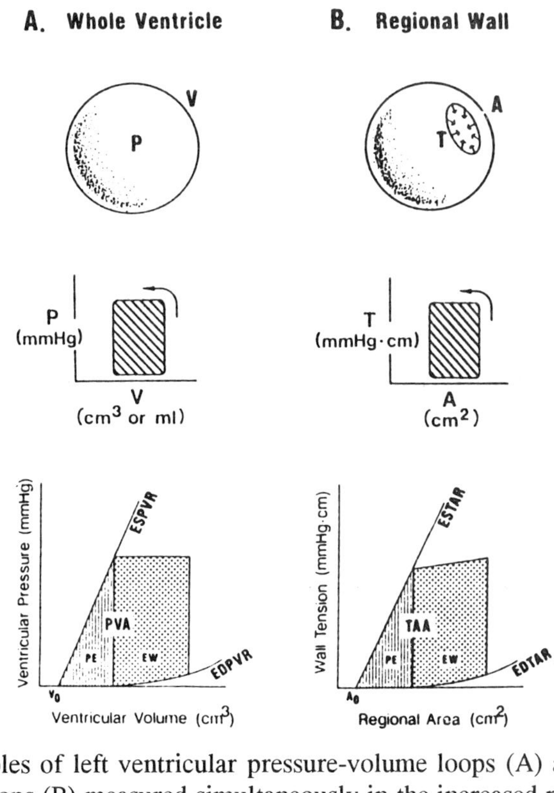

Fig. 2. Examples of left ventricular pressure-volume loops (A) and wall tension-regional area loops (B) measured simultaneously in the increased preload run. C to E depict correlations between regional myocardial oxygen consumption (rVO_2) and percent area shrinkage (C), regional work (D), and TAA (E). All three variables correlated positively with rVO_2. The inner pair of dashed curves indicates 95% confidence limit of the regression line, while the outer pair of dashed lines indicates 95% confidence limit of sampled data. Reprinted with permission from ref. 14.

volume run, regional area shrinkage decreased gradually as end-diastolic and end-systolic volumes were increased, resulting in a negative correlation with regional VO_2. In contrast, both regional work and TAA progressively increased, resulting in linear, positive correlations with regional VO_2. In the contraction mode run, neither regional area shrinkage nor regional work correlated significantly with regional VO_2. In contrast, TAA correlated linearly with regional VO_2. Interestingly, the slope and regional VO_2 intercept of the regional VO_2-TAA relation were

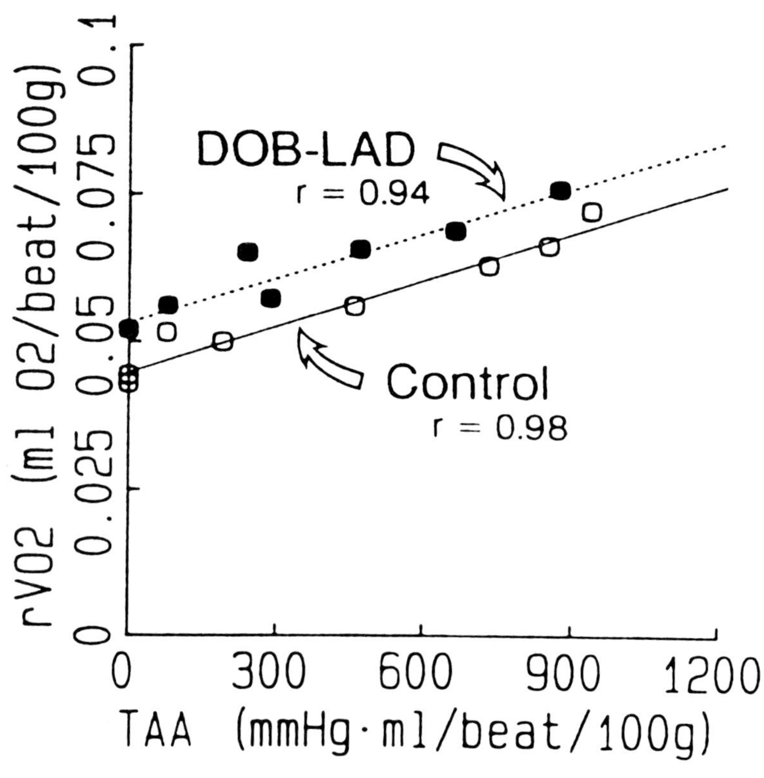

Fig. 3. Examples of left ventricular pressure-volume loops (A) and wall tension-regional area loops (B) measured in the decreased afterload run. C to E depict correlations between regional myocardial oxygen consumption (rVO_2) and percent area shrinkage (C), regional work (D), and TAA (E). Area shrinkage negatively correlated with rVO_2 and regional work did not show a significant correlation, whereas TAA positively correlated with rVO_2. Pairs of dashed curves as in Figure 2. Reprinted with permission from ref. 14.

remarkably consistent among the four loading protocols despite the considerable differences in preload and afterload conditions. The slope and regional VO_2 intercept of the regional VO_2-TAA relation in all 18 runs averaged $2.22 \times 10^{-5} \pm 0.78 \times 10^{-5}$ ml $O_2 \cdot mmHg^{-1} \cdot ml^{-1}$ and 0.037 ± 0.009 ml $O_2 \cdot beat^{-1} \cdot 100g^{-1}$, respectively.

Regional Contractility Protocol

Figure 4 shows the regional VO_2-TAA relation before and during infusion of dobutamine into the left anterior descending coronary artery in a representative heart. In this heart, the slope of the ESTAR, regional E_{max} of the anterior wall, increased by 42.1% with dobutamine, whereas global E_{max} increased only slightly, by 4.5%, indicating that the ESTAR sensitively reflects changes in regional myocardial contractility. The regional VO_2-TAA relation was highly linear during both the control period and dobutamine infusion. With dobutamine, it shifted upward in a parallel manner. This parallel upward shift with dobutamine is similar to the findings for the VO_2-PVA relation obtained in the whole left ventricle (6,7). Thus, the regional VO_2-TAA relation sensitively reflects regional changes in myocardial contractility in the same manner as that for the whole ventricular VO_2-PVA relation.

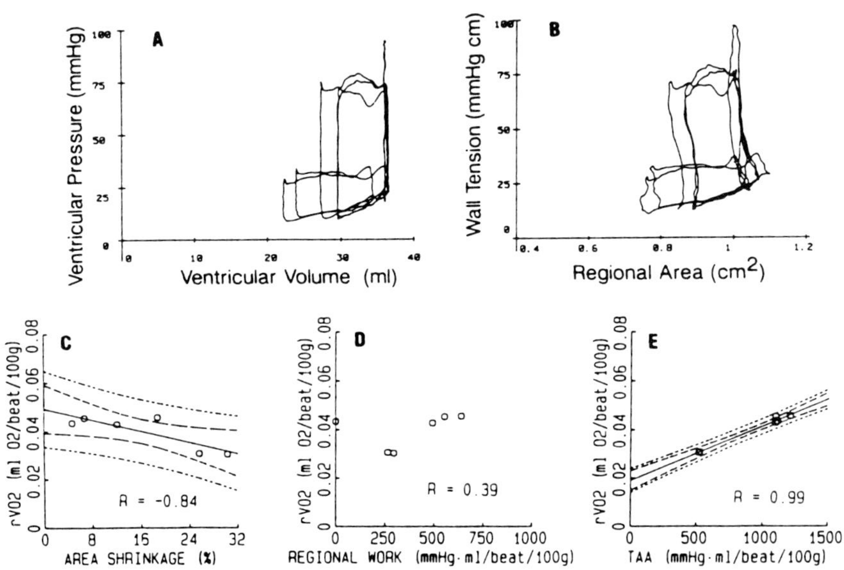

Fig. 4. The relationship between regional myocardial oxygen consumption (rVO_2) and tension-area area (TAA) in the anterior wall region during control period and dobutamine (DOB) infusion into the left anterior descending coronary artery (LAD). During enhancement of regional contractility with DOB, the linear relation between rVO_2 and TAA shifted upward in a parallel manner. Reprinted with permission from ref. 13.

DISCUSSION

Conventional Indexes of Regional VO_2

Many variables have been proposed as a mechanical index of VO_2 including pressure-rate product (17), tension-time index (18), force-time integral (19), and pressure-work index (20). However, all these are global rather than regional mechanical variables, and therefore reflect whole ventricular VO_2 rather than regional VO_2. Recently, Gayheart et. al. (15) observed a significant, but modest correlation (r=0.70) between regional VO_2 and segment shortening during regional myocardial dysfunction induced by intracoronary lidocaine infusion. An interesting observation in their study was that VO_2 in the anterior wall region decreased by only 33% when systolic segment shortening of the region was completely suppressed by intracoronary lidocaine infusion. From the viewpoint of TAA, this disproportionately small reduction of regional VO_2 despite complete akinesis (zero external work) could be explained by the possibility that oxygen was consumed by the remaining potential energy portion of TAA (Figure 1B) or by nonmechanical activities (6) which are reflected by the regional VO_2-intercept of the regional VO_2-TAA relation (Panel E in Figures 2 and 3).

TAA as a Measure of Regional Total Mechanical Energy

TAA showed a strong, positive correlation with regional VO_2 under any set of loading conditions and contractility in the present study. This strong correlation is quite reasonable because TAA has been derived by analogy to PVA, which is considered the best correlate of VO_2 (6,9,21). Because PVA has been considered to represent the total mechanical energy expenditure of the global ventricle during one cardiac contraction (6), TAA can also be considered as a measure of total mechanical energy expenditure of the region during one myocardial contraction. In addition, because the dimensions of both regional VO_2 and TAA can be converted to those of energy (J) according to the equations, 1 ml O_2 = 20 J and 1 mmHg·ml = 1.33×10^{-4} J, the reciprocal of the slope of the regional VO_2-TAA relation is considered to be the chemomechanical energy conversion efficiency [i.e. contractile efficiency (6,8)] of the region from the oxygen utilized for mechanical contraction (chemical energy) to TAA (total mechanical energy). In fact, the slope of the regional VO_2-TAA relation was remarkably consistent regardless of left ventricular loading conditions and close to the reported value for the slope of the whole ventricular VO_2-PVA relation ($1.5\text{-}2.2 \times 10^{-5}$ ml $O_2 \cdot mmHg^{-1} \cdot ml^{-1}$) (6-9). The close agreement of the slope of the regional VO_2-TAA and VO_2-PVA relations sug-gests that the contractile efficiency is similar between TAA (regional total mechanical energy) and PVA (ventricular total mechanical energy). On the other hand, the regional VO_2 intercept at zero TAA is considered to reflect oxygen consumption utilized for nonmechanical activities such as excitation-contraction coupling and basal metabolism (6,8), because dobutamine increased only the regional VO_2 intercept without changing the slope of the regional VO_2-TAA relation. Thus, the theoretical background and experimental findings indicate that there is a tight coupling between regional VO_2 per beat and regional total mechanical energy generated by a contraction.

Future Direction

At present, it is not easy to directly measure TAA and regional VO_2 in clinical settings. However, recent advances in technology have allowed noninvasive measurements of both segment strain by magnetic resonance imaging using a myocardial tagging technique (22) and regional VO_2 by positron emission tomography (23). With these new methods, it might become possible to measure TAA noninvasively in a clinical setting. As a more practical alternative to the T-A relation, Nakano et. al (24) have proposed the relation between mean wall stress and "natural logarithm of the reciprocal of wall thickness". With this method, an assessment of the energetic equivalent of TAA would be possible using echocardiography in a clinical setting.

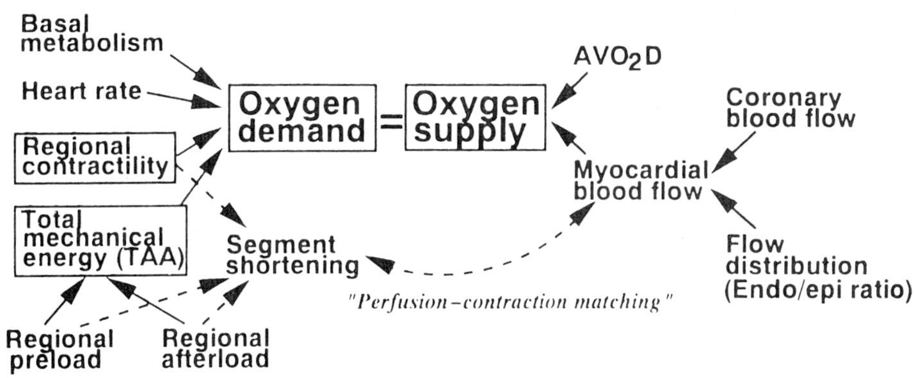

Fig. 5. Determinants of regional myocardial oxygen demand and supply. Regional myocardial oxygen demand is primarily determined by regional total mechanical energy (i.e., TAA), regional myocardial contractility, heart rate and basal metabolism. Regional myocardial oxygen supply is determined by regional myocardial blood flow and arteriovenous oxygen content difference (AVO_2D).

CONCLUSION

Conventional variables of regional contractile function, such as the amount of myocardial shortening and regional work, cannot predict regional VO_2 because of load-dependence. In contrast, TAA is a reliable predictor of regional VO_2 per beat regardless of ventricular loading conditions. In addition, the regional VO_2-TAA relation sensitively reflects contractility-dependent alteration of regional myocardial energetics. The reciprocal of the slope of the regional VO_2-TAA relation may indicate regional contractile efficiency, while the regional VO_2 intercept may reflect regional nonmechanical energy expenditure utilized for E-C coupling and basal metabolism. These findings indicate that regional VO_2 is tightly coupled with regional total mechanical energy represented by TAA. Using the concept of TAA, we may obtain better insight into physiology and pathophysiology of regional myocardial mechanics and energetics.

ACKNOWLEDGEMENT

Supported in part by a Research Grant for Cardiovascular Diseases from the Ministry of Health and Welfare of Japan (4A-1, 4C-4).

REFERENCES

1. Theroux, P., Franklin, D., Ross, J., Jr., Kemper, W.S. Regional myocardial function during acute coronary artery occlusion and its modification by pharmacologic agents in the dog. Circ Res 35:896-908, 1974.

2. Weintraub, W.S., Hattori, S., Agarwal, J.B., Bodenheimer, M.M., Banka, V.S., Helfant, R.H. The relationship between myocardial blood flow and contraction by myocardial layer in the canine left ventricle during ischemia. Circ Res 48:430-438, 1981.

3. Gallagher, K.P., Matsuzaki, M., Osakada, G., Kemper, W.S., Ross, J., Jr., Effect of exercise on the relationship between myocardial blood flow and systolic wall thickening in dogs with acute coronary stenosis. Circ Res 52:716-729, 1983.

4. Forrester, J.S., Wyatt, H.L., Da Luz, P.L., Tyberg, J.V., Diamond, G.A., Swan, H.J.C. Functional significance of regional ischemic contraction abnormalities. Circulation 54:64-70, 1976.

5. Ross, J., Jr. Myocardial perfusion-contraction matching. Implications for coronary heart disease and hibernation. Circulation 83:1076-1083, 1991.

6. Suga, H. Ventricular energetics. Physiol Rev 70:247-277, 1990.

7. Suga, H., Hisano, R., Goto, Y., Yamada, O., Igarashi, Y. Effect of positive inotropic agents on the relation between oxygen consumption and systolic pressure volume area in canine left ventricle. Circ Res 53:306-318, 1983.

8. Goto, Y., Slinker, B.K., LeWinter, M.M. Decreased contractile efficiency and increased nonmechanical energy cost in hyperthyroid rabbit heart: relation between O_2 consumption and systolic pressure-volume area or force-time integral. Circ Res 66:999-1011, 1990.

9. Schipke, J.D., Burkhoff, D., Kass, D.A., Alexander, J., Schaefer, J., Sagawa, K. Hemdynamic dependence of myocardial oxygen consumption indexes. Am J Physiol 258:H1281-H1291, 1990.

10. Goto, Y., Suga, H., Yamada, O., Igarashi, Y., Saito, M., Hiramori, K. Left ventricular regional work from wall tension-area loop in canine heart. Am J Physiol 250:H151-H158, 1986.

11. Goto, Y., Igarashi, Y., Yasumura, Y., Nozawa, T., Futaki, S., Hiramori, K., Suga, H. Integrated regional work equals total left ventricular work in regionally ischemic canine heart. Am J Physiol 254:H894-H904, 1988.

12. Goto, Y., Igarashi, Y., Yamada, O., Hiramori, K., Suga, H. Hyperkinesis without the Frank-Starling mechanism in a nonischemic region of acutely ischemic excised canine heart. Circulation 77:468-477, 1988.

13. Goto, Y., Suga H. Left ventricular regional mechanics and energetics assessed in the wall tension-regional area diagram. In Recent Progress in Failing Heart Syndrome (eds. Sasayama, S., Suga, H.), Springer-Verlag, Tokyo, pp. 117,150, 1991.

14. Goto, Y., Futaki, S., Kawaguchi, O., Hata, K., Takasago, T., Saeki, A., Nishioka, T., Taylor, T.W., Suga, H. Coupling between regional myocardial oxygen consumption and contraction under altered preload and afterload. J Am Coll Cardiol 21:1522-1531, 1993.

15. Gayheart, P.A., Vinten-Johansen, J., Johnston, W.E., Hester, T.O., Cordell, A.R. Oxygen requirements of the dyskinetic myocardial segment. Am J Physiol 257:H1184-H1191, 1989.

16. Goto, Y., Futaki, S., Ohgoshi, Y., Yaku, H., Kawaguchi, O., Suga, H. Assessment of left ventricular regional work under ischemia. Frontiers Med Biol Engng 2:201-205, 1990.

9. Energetics During Ventricular Fibrillation

Hideo Kusuoka and Eduardo Marban

INTRODUCTION

Ventricular fibrillation (VF) is a common, often fatal arrhythmia frequently complicating acute myocardial infarction or heart failure. The pathophysiology of ventricular fibrillation has attracted much interest not only from the viewpoint of basic electrophysiology, but also that of clinical cardiology. Despite the importance of this arrhythmia, little is known about the cellular metabolism of the heart when it is subjected to ventricular fibrillation. In this study, phosphorus-31 magnetic resonance spectroscopy (P-NMR) was applied to isolated perfused ferret hearts to investigate the alterations in myocardial energetics induced by VF.

CHANGES IN ENERGY METABOLISM DURING VENTRICULAR FIBRILLATION AND INTRACELLULAR CALCIUM OVERLOAD

Metabolic Changes During VF Induced by Digitalis Intoxication

In calcium-overload states, such as digitalis intoxications, oscillation of intracellular free Ca^{++} concentration ($[Ca^{++}]_i$) have been implicated as the cause of VF (1). When strophanthidin (20μM) was added during pacing at 3 Hz, the positive inotropic effect soon gave way to a decrease in developed pressure. The decrease in force was accompanied by changes similar to those in hypoxia, i.e., an increase in intramyocardial concentration of inorganic phosphate ([Pi]), a decrease in phosphocreatine ([PCr]), and a slight acidosis. The rhythm changed to VF after 12-25 minutes. This change was initially accompanied by further metabolic deterioration (Fig. 1B); Pi increased considerably, with a corresponding decrease in PCr. There was also a decrease in [ATP] of almost 50% and slight acidosis. However, these metabolites reached steady state within 12-18 minutes of the onset of VF. The spectrum shown in Fig. 1C was taken 18 minutes later, but is almost identical to that in Fig. 1B.

Effect of Ryanodine on Energy Metabolism During VF

The changes in P-NMR during VF induced by digitalis intoxication, such as the increase in Pi and the decrease in PCr, are very similar to those observed during hypoxia, indicating imbalance between energy supply and energy demand. Thus, two mechanisms need to be considered as the cause for these changes; one is an increase in energy demand induced by VF, and the second is a limitation of energy supply induced by calcium overload.

To dissect these two possible mechanisms, we applied ryanodine to the hearts during VF. Ryanodine is an alkaloid that interferes with calcium release from the sarcoplasmic reticulum and inhibits calcium oscillations (2). When ryanodine (100nM) was added to the perfusate in the continuing presence of strophanthidin, a steady-state effect was reached within 6 minutes; ryanodine abolished the pressure oscillations but not the voltage oscillations. This striking dissociation

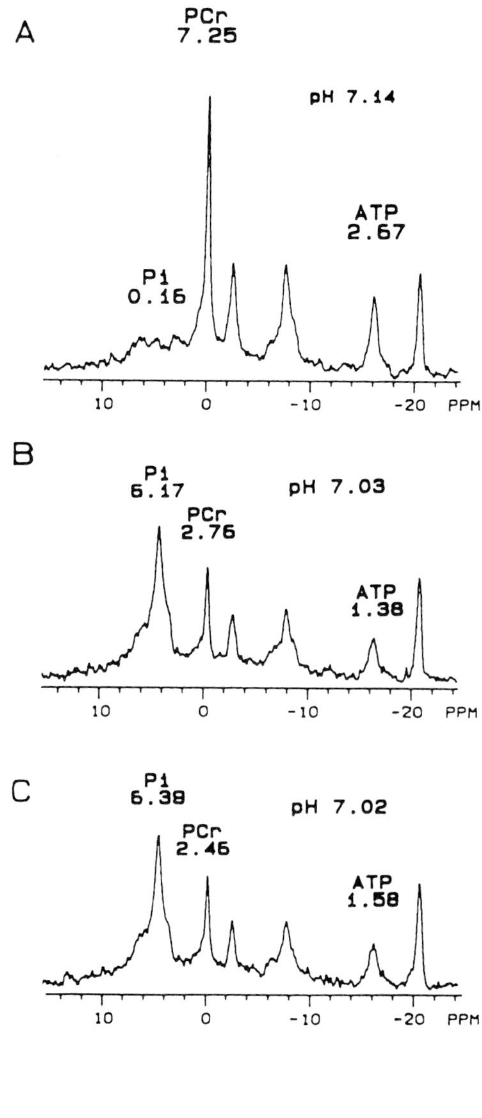

Fig 1. Myocardial metabolites reach steady state during VF. P-31 NMR spectra (6-minutes acquisition) from a representative heart are shown during perfusion with control solution (Panel A), and at two times during VF (12-18 minutes, Panel B; 30-36 minutes, Panel C). The identity and concentration in μ moles/g wet wt of each metabolites is indicated above the corresponding peak, except for pH, which is displayed in the upper right-hand corner of each spectrum. Reprinted with permission from ref. 3.

between the mechanical and electrical processes indicates that oscillations of intracellular calcium are not required to sustain VF.

The contribution of calcium cycling to the metabolic deterioration was tested by comparing P-NMR spectra before and during exposure to ryanodine. Figure 2B shows the spectrum obtained during the exposure to ryanodine. When compared with the spectrum obtained during VF, but before application of ryanodine (Fig. 2A), there is a marked increase in [PCr] and a concomitant fall in [Pi]. [ATP] and pH are little affected.

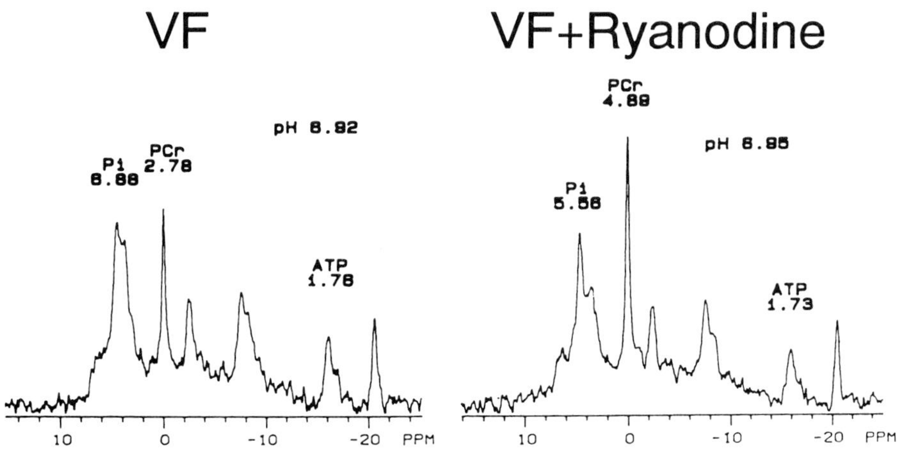

Fig. 2. Effect of ryanodine on myocardial metabolites in one heart. Individual P-31 NMR spectra are shown before (left) and during 12-18 minutes with addition of ryanodine (100nM) to the perfusate (right). Reprinted with permission from ref. 3.

Figure 3 summarizes the effects of ryanodine on these metabolites during VF. The improvement in [Pi] and [PCr], and in their ratio ([Pi]/[PCr]) with ryanodine was a consistent finding in all experiments ($P<0.05$). However, the restoration toward control values is incomplete. After ryanodine, the hearts exhibited a slightly higher mean pressure than before, suggesting that the energy demand did not decrease after ryanodine, and that the beneficial effect of ryanodine is not due to mechanical unloading. These results indicate that oscillations of $[Ca^{++}]_i$ may contribute importantly not only to energy consumption but also to metabolic deterioration (3).

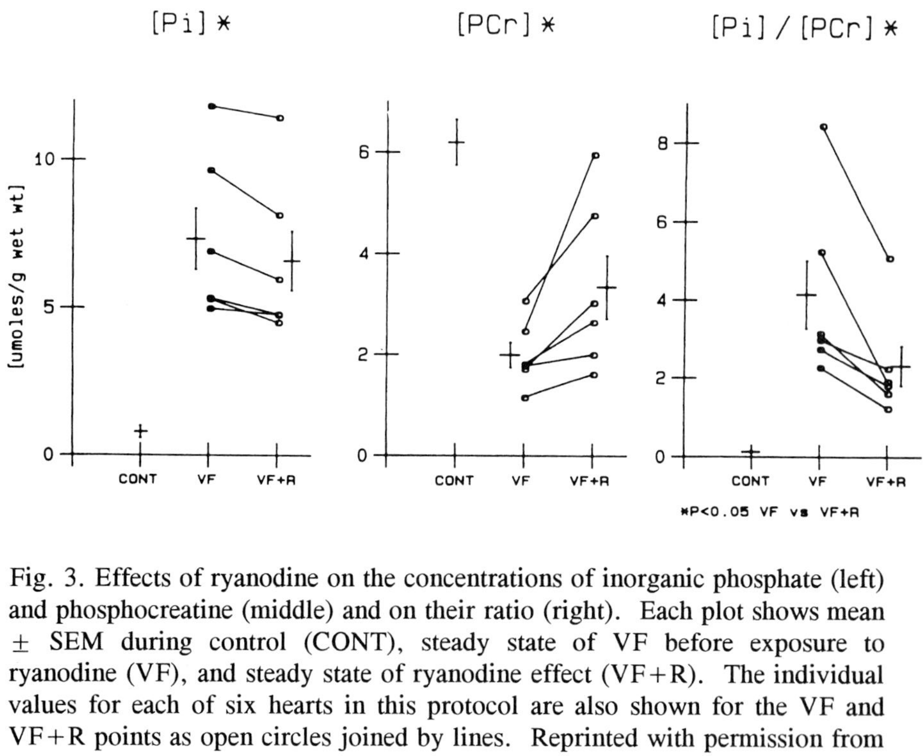

Fig. 3. Effects of ryanodine on the concentrations of inorganic phosphate (left) and phosphocreatine (middle) and on their ratio (right). Each plot shows mean ± SEM during control (CONT), steady state of VF before exposure to ryanodine (VF), and steady state of ryanodine effect (VF+R). The individual values for each of six hearts in this protocol are also shown for the VF and VF+R points as open circles joined by lines. Reprinted with permission from ref. 3.

Metabolic Changes and Calcium Overload During VF Induced by Electrical Stimulation

In contrast to digitalis-induced VF, metabolic changes observed by P-NMR were mild in electrically-induced VF; VF induced by repetitive ventricular stimulation at 10-30 Hz led to only a small increase in [Pi] and a small decrease in [PCr]. Nevertheless, $[Ca^{++}]_i$ increased markedly during electrically-induced VF; the time-averaged $[Ca^{++}]_i$ increased about five-fold during VF (4).

Energy Flux During Ventricular Fibrillation

The difference of the changes in energy-related phosphates between two types of VF suggest that the cellular energetics in the hearts differed based on the degree of intracellular calcium overload.

Cytosolic ATP consumption occurs at two principal sites, the contractile elements and ion-transport proteins, including the sarcoplasmic reticulum calcium pump and sarcolemmal Na^+,K^+-ATPase. Failure to improve energy metabolism significantly during VF by the simple abolition of force generation suggests that the increase in energy requirement during VF is mainly attributable to the metabolic cost of calcium cycling. On the other hand, calcium overload itself has been proposed to decrease the efficiency of ATP synthesis by mitochondria (5). In either case, a mismatch of energy supply and demand would result. Thus, two possible hypotheses could explain the changes in energy metabolism during VF. The first invokes a failure of ATP synthesis in mitochondria that is due to calcium overload. The second hypothesis postulates an energy imbalance that is primarily due to an increase in consumption.

Creatine Shuttle

In myocardium, mitochondria are the major site of energy production through the process of oxidative phosphorylation. ATP produced in mitochondria is not transferred directly to the cytosol, where ATP-consuming reactions occur, but rather is effectively transported by the "creatine shuttle" (6): Creatine kinase (CK) coupled to ATP synthesis in mitochondria transmits the energy in ATP to PCr, PCr diffuses into the cytosol, and ATP is then resynthesized by cytosolic CK associated with the myofilaments. A failure of energy transfer by CK during VF could, in principle, contribute to an insufficient energy supply.

The magnetization transfer method in NMR allows determination of the flux of specific reactions (7,8). Creatine kinase governs the reaction between PCr and ATP. ATP is synthesized with ADP and inorganic phosphate. Thus, the rate whereby Pi is incorporated into ATP and the flux catalyzed by CK reaction are both measurable by P-NMR spectroscopy, enabling the quantification of the ATP synthesis rate (5) and energy transfer capacity by CK (6) in the myocardium.

Energy Flux During VF Induced By Digitalis Intoxication

To elucidate the disorder of energy metabolism in myocardium during VF, we measured energy transfer capacity and the ATP synthesis rate in myocardium. During VF induced by digitalis intoxication, the ATP synthesis rate significantly increased to levels about three fold higher than control (Fig.4 left). In contrast, the calculated flux from phosphocreatine to ATP, that is, the energy transfer capacity was significantly lower than that in control, but still larger than the ATP synthesis rate (Fig. 4 middle). As illustrated in the right panel of Fig. 4, oxygen consumption increased by only 16%. These results indicate that, during VF induced by digitalis intoxication, energy production increases compared with regular rhythm, but the increase in energy consumption overwhelms the increase in energy production.

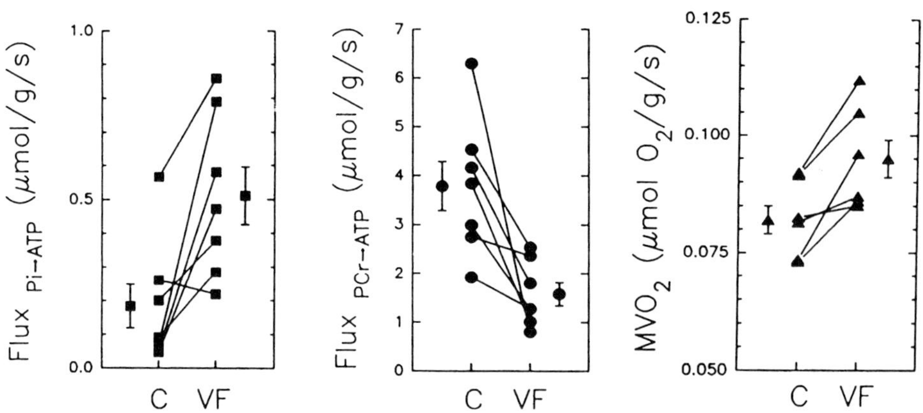

Fig. 4. The changes in the fluxes in the ATP synthesis rate ($Flux_{Pi \rightarrow ATP}$) and the energy transfer capacity ($Flux_{PCr \rightarrow ATP}$), and oxygen consumption (MVO_2) during VF induced by digitalis intoxication. C, the control condition. Reprinted with permission from ref. 9.

Energy Flux During Electrically-Induced VF

In contrast to the marked energetic deterioration during VF induced by digitalis, the changes in the phosphorus compounds during electrically-induced VF were small. Furthermore, the ATP synthesis rate increased, but it was not significant (Fig. 5). The flux from phosphocreatine to ATP was maintained at levels as high as control values (Fig.5, middle). In contrast to VF induced by digitalis intoxication, oxygen consumption during electrically induced VF did not show consistent changes (Fig.5). These results reveal that, during VF induced by electrical stimulation, energy flux was maintained as well as under control conditions.

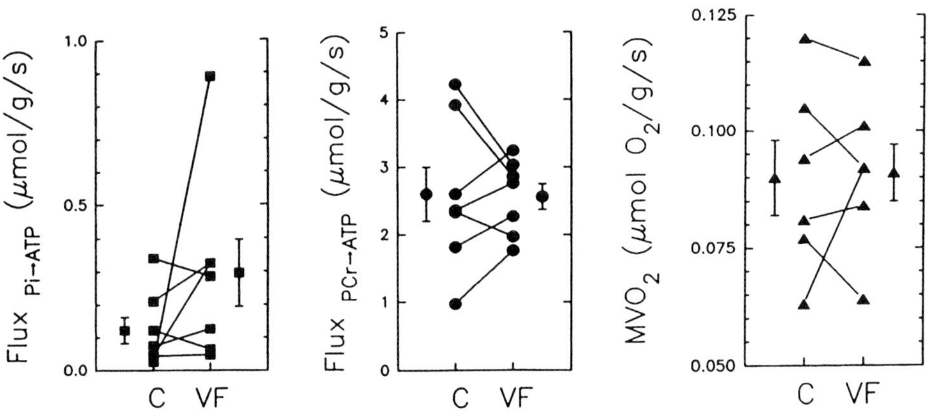

Fig. 5. The changes in the fluxes in the ATP synthesis rate ($Flux_{Pi \rightarrow ATP}$) and the energy transfer capacity ($Flux_{PCr \rightarrow ATP}$), and oxygen consumption (MVO_2) during VF induced by electrical stimulation. C, the control condition. Reprinted with permission from ref. 9.

DIFFERENT EFFECTS OF VF ON ENERGY METABOLISM

Our results reveal important differences in energy metabolism during VF, depending on the specific method used to elicit VF. Fibrillation triggered by digitalis intoxication produced a metabolic picture similar to hypoxia, where despite an increase in the ATP synthesis rate, the capacity of energy transfer decreased but still remained high enough to overwhelm the ATP synthesis rate. These results suggest a deterioration in energy production, as well as a marked increase in energy demand, as the major cause of the hypoxia-like changes during VF induced by digitalis. In contrast, energy supply during VF induced by electrical stimulation was at least as high as that found under control conditions. Nevertheless, the modest increase in [Pi], coupled with a decrease in [PCr], suggests that the energy supply did not quite suffice to maintain steady-state metabolite concentrations equal to those in the control condition, presumably because of a disproportionate increase in energy demand.

Figure 6 shows the relation between the ATP synthesis rate and the left ventricular pressure during VF. In VF induced by electrical stimulation, ATP synthesis rate is independent of LV pressure. In contrast, during VF induced by digitalis, ATP synthesis rate negatively correlates with LV pressure ($r=-0.95$, $P<0.02$), indicating that the severity of intracellular calcium overload limits ATP synthesis rate. Neither the oxygen consumption rate during digitalis-induced VF nor that during electrically-induced VF correlated with the ATP synthesis rate.

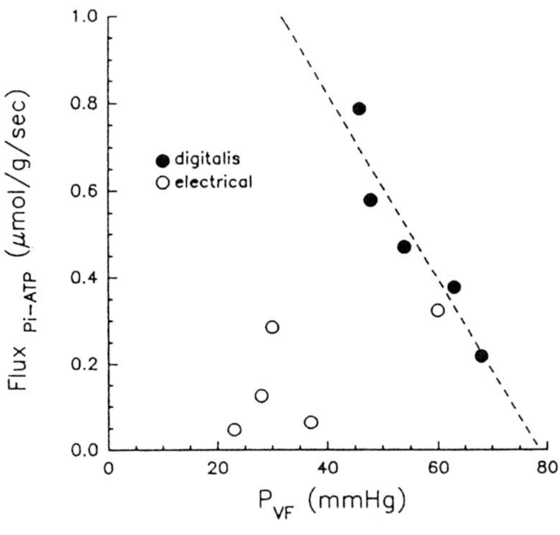

Fig. 6. The relation between left ventricular pressure (P_{VF}) and ATP synthesis rate ($Flux_{Pi \rightarrow ATP}$) during VF induced by digitalis intoxication (●) or by electrical stimulation (○). The dash line represents the best-fit regression line for the data during digitalis-induced VF: $Flux_{Pi \rightarrow ATP}=-0.021p_{VF}+1.68$. Reprinted with permission from ref. 9.

Although the ATP synthesis rate and the oxygen consumption were not measured simultaneously, it is possible to estimate the apparent P/O ratio using the regression between the ATP synthesis rate or myocardial oxygen consumption and ventricular work load. The P/O ratio estimated from our experimental data is very high (7.8 -8.0), as observed in moderately loaded hearts (7), suggesting high futile cycling of ATP. Since there is no rigorous measure to quantify the work load of hearts during VF, it is difficult to estimate the P/O ratio during VF. If individual groups are averaged to estimate the variables, the P/O ratio during VF induced by electrically stimulation dropped to 3.2, whereas the apparent P/O ratio during digitalis-induced VF ranged from 9.0 to 2.3 depending on the left ventricular pressure during VF. Taken together with the changes in P-NMR spectra during VF, these results indicate that energy was efficiently generated by oxidative phosphorylation during electrically induced VF, whereas ATP production during digitalis-induced VF was restricted.

CONCLUSIONS

These results indicate that the rundown in energy metabolism during VF induced by digitalis was mainly attributable to a limitation of energy production through oxidative phosphorylation as well as to a marked increase in energy consumption. In contrast, myocardial energy generation remained unimpaired during VF induced by electrical stimulation. Intracellular calcium overload is more severe during VF induced by digitalis than during electrically induced VF (1); severe calcium overload would be expected to compromise the capacity for energy generation by mitochondria. Thus, we propose that known differences in cellular calcium loading underlie the discrepant energetic patterns of the two types of VF (9).

REFERENCES

1. Kihara, Y., Morgan, J.P. Intracellular calcium and ventricular fibrillation: Studies in the aequorin-loaded isovolumic ferret heart. Circ Res 68:1378-1389, 1991.

2. Sutko, J.L., Kenyon, J.L. Ryanodine modification of cardiac muscle responses in potassium-free solutions: Evidence for inhibition of sarcoplasmic reticulum release. J Gen Physiol 82:385-404, 1983.

3. Kusuoka, H., Jacobus, W.E., Marban, E. Calcium oscillations in digitalis-induced ventricular fibrillation: Pathogenetic role and metabolic consequences in isolated ferret hearts. Circ Res 62:609-619, 1988.

4. Koretsune, Y., Marban, E. Cell calcium in the pathophysiology of ventricular fibrillation and in the pathogenesis of postarrhythmic contractile dysfunction. Circulation 80:369-379, 1989.

5. Jacobus, W.E., Tiozzo, R., Lugli, G., Lehninger, A.L., Carafoli, E. Aspects of energy-linked calcium accumulation by rat heart mitochondria. J Biol Chem 250:7863-7870, 1975.

6. Jacobus, W.E. Respiratory control and the integration of heart high-energy phosphate metabolism by mitochondrial creatine kinase. Annu Rev Physiol 47:707-725, 1985.

7. Kingsley-Hickman, P.B., Sako, E.Y., Mohanakrishnan P, Robitaille PML, From AHL, Foker JE, Ugurbil K. ^{31}P NMR studies of ATP synthesis and hydrolysis kinetics in the intact myocardium. Biochemistry 26:7501-7510, 1987.

8. Neubauers, S., Hamman, B.L., Perry, S.B., Bittl, J.A., Ingwall, J.S. Velocity of the creatine kinase reaction decreases in postischemic myocardium: A ^{31}P-NMR magnetization transfer study of the isolated ferret heart. Circ Res 63:1-15, 1988.

9. Kusuoka, H., Chacko, V.P., Marban, E. Myocardial energetics during ventricular fibrillation investigated by magnetization transfer nuclear magnetic resonance spectroscopy. Circ Res 71:1111-1122, 1992.

10. Influence of Preload on Non-Mechanical VO_2 Assessed with 2,3-Butanedione Monoxime

Martin M. LeWinter, Akihiro Higashiyama, Hitoshi Yaku, and Matthew W. Watkins

INTRODUCTION

In isolated heart muscle, muscle length has a positive effect on both level of activation and energy consumption for excitation-contraction (EC) coupling (1-8). Both an increase in myofilament Ca^{++} affinity and an increase in total systolic Ca^{++} release have been identified as mechanisms of length-dependence of activation (1,3,5-7). With respect to energy consumption, only an increase in Ca^{++} release would be anticipated to increase energy consumption for E-C coupling, as a result of the requirement for increased Ca^{++} reuptake by the sarcoplasmic reticulum, an energy requiring process. In contrast to the aforementioned results in isolated muscle, in studies performed in the beating heart, non-mechanical O_2 consumption (VO_2) has not been shown to be influenced by preload (9). Thus, with respect to energetics, there is an apparent discrepancy between the results obtained in isolated heart muscle as opposed to the whole heart.

As indicated above, in whole heart studies non-mechanical VO_2 has been employed as a measure of energy consumed for the combination of EC coupling and basal metabolism (10,11). Non-mechanical VO_2 is the oxygen consumption present under mechanically unloaded conditions, which is typically obtained by reducing ventricular volume to a point at which there is no positive pressure generation. Since basal VO_2 is not influenced by preload (12), lack of a change in unloaded O_2 as a function of preload implies that energy consumption for EC coupling is also unaffected by preload. However, it is intuitively obvious that the mechanically unloaded VO_2 must contain some energy consumption for crossbridge cycling in addition to EC coupling and basal metabolism. This is because the ventricle visibly deforms and changes in shape with each contraction under unloaded conditions and, in addition, there are pressure fluctuations within the negative range in the unloaded ventricle. It is therefore possible that assessment of EC coupling energy consumption based on changes in unloaded VO_2 is misleading because of this uncertain contribution of crossbridge cycling energy consumption. For this reason, we recently adapted a new technique for assessing non-mechanical VO_2 in the beating left ventricle using the drug 2,3-butanedione monoxime (BDM) (13). This method is based on the report of Alpert et al.(8) indicating that BDM is a virtually myofilament selective negative inotrope in isolated rabbit heart muscle at perfusate concentrations less than about 6mM. Alpert et al.(8) exploited this feature of BDM to partition initial heat into tension-dependent (crossbridge cycling) and tension-independent (EC coupling) components in myothermal experiments performed in isometrically contracting rabbit heart muscle. In one variation of their method, BDM was administered at gradually increasing concentrations up to 6mM and the linear relation between initial heat and tension time integral extrapolated to the initial heat axis as an estimate of tension-independent heat. Our adaptation of their method employs this approach. Specifically, BDM is administered in increasing concentrations up to 5mM in the perfusate of the red blood cell-perfused, isovolumically contracting rabbit heart preparation (13). The linear relation between VO_2 and force time integral obtained under control conditions and then with increasing concentrations of BDM is extrapolated to the VO_2 axis to provide an estimate of non-mechanical VO_2. Because of the myofilament selectivity of BDM, this estimate of non-mechanical VO_2 should contain only VO_2 used for EC coupling and basal metabolism, without contamination by crossbridge cycling energy consumption. In the present study, we applied the BDM method in the red blood cell-perfused, isolated rabbit ventricle to determine if variations in preload within the physiologic range influence non-mechanical VO_2. The issue of energy consumption for EC coupling in the beating heart is pertinent not just to length dependent activation but also to the VO_2-pressure-volume area (PVA) concept. Use of the VO_2-PVA relation to assess contractile efficiency requires an assumption that energy used for EC coupling does not vary as a function of PVA. In developing the VO_2-PVA relation, ventricular volume typically is altered in order to vary PVA.

METHODS

Heart Preparation

Our isolated rabbit heart preparation perfused with bovine red blood cells suspended in Krebs-Henseleit buffer has been described in detail elsewhere (14). Briefly, 16 adult New Zealand white rabbits [body weight 3.29 ± 0.25 (SD) kg] were anesthetized with ketamine hydrochloride (20 mg/kg IM) and xylazine (1 mg/kg IM) after medication with fentanyl (0.044 mg/kg IM) and droperidol (2.2 mg/kg IM). After tracheotomy, ventilation was maintained with a Harvard respirator. The chest was then opened mid-sternally, the pericardium widely incised, and both pulmonary hili and the superior and inferior venae cavae ligated simultaneously. Cannulae were introduced into the ascending aorta and right ventricle (RV), and the heart removed from the chest. The left ventricular (LV) apex was vented, the left atrium opened widely, and a thin latex balloon (unstressed volume 2-3ml) was inserted into the LV via the mitral orifice. The balloon lumen was connected to a graduated syringe and pressure transducer (Gould P23XL). The RV was vented to air and collapsed by continuous drainage. Coronary perfusion pressure was maintained by a pressurized chamber connected to compressed air and was monitored continuously. Coronary blood flow was measured by timed collections into a graduated cylinder of the coronary venous drainage from the RV. LV thebesian flow was neglected, because it is a very small fraction of total flow (10). Arteriovenous O_2 content difference was measured continuously with an AVOX system (A-VOX, San Antonio, TX) that was calibrated with a Lex-02-Con O_2 content analyzer (LEX-02-Con, Waltham, MA). Heart rate was kept constant (average 162 ± 17 beats/min) by left atrial pacing and coronary perfusion pressure was maintained at approximately 80mmHg throughout each experiment.

Preparation of Perfusate

The perfusate consisted of bovine red blood cells suspended in Krebs-Henseleit buffer. The procedure for treating and storing the red blood cells has been described in detail elsewhere (14). The buffer had the following composition (in mM): 108.0 NaCl, 4.0 KCl, 2.5 $CaCl_2$, 1.5 KH_2P0_4, 25.0 $NaHCO_3$, 11.0 dextrose, and 10.0 Na-pyruvate (all from Sigma Chemical). Before each experiment, the red cells were suspended to a final hematocrit of 35% in the Krebs-Henseleit buffer and equilibrated with 98% O_2-2% CO_2 to achieve a PO_2 > 100 Torr and PCO_2 of approximately 40 Torr. pH was adjusted to 7.35-7.45 by addition of NaOH. Perfusate temperature was maintained between 35-37°C with water jackets around an oxygenator and a pressurized chamber in the arterial tube. The perfusate was not recirculated.

Experimental Protocol

LV pressure, coronary blood flow and arteriovenous O_2 difference were measured during steady-state isovolumic contractions at a volume at which LV peak isovolumic pressure was 0 mmHg(unloaded volume, V_0). Then, LV volume was increased until a point at which end-diastolic pressure was 10-12 mmHg or unstable contractions appeared. (In 9 of 16 hearts, an end-diastolic pressure of 10-12 mmHg could not be achieved for this reason.) The latter volume is referred to subsequently as high volume (V_H). Once V_H was achieved, physiologic variables were measured once again. LV volume was then decreased to a point at which LV peak pressure was 60-70% of that present at V_H, and measurements were repeated. We refer to the latter as low volume (V_L). With each change in LV volume, we waited 2-3 minutes to obtain stable conditions. Finally, LV volume was decreased to V_0 once more.

As previously described (13), BDM was then infused at 4 or 5 perfusate concentrations less than or equal to 5 mM. Once stable conditions were obtained at each [BDM] measurements were repeated at V_H and V_L.

Following this, the heart was arrested by rapidly infusing 20% KCL solution (0.2 ml/min) followed by a slower infusion at 0.03 ml/min. Coronary blood flow and arteriovenous 0_2 difference were measured at V_0, 15 minutes after cardiac arrest. This same measurement was then repeated at V_H. After the LV volume change, we waited 2-3 minutes. At the end of each experiment, the LV (free wall + septum) and the RV (free wall only) were weighed. LV weight was 5.64 ± 0.77g and RV weight was 1.59 ± 0.33g.

Data Analysis

Physiologic variables were collected with a personal computer (Gateway 2000, North Sioux City, SD) at 5 ms sampling intervals and stored on hard disk for off-line analysis.

Total VO_2 per minute was calculated as coronary flow (ml/min)x arteriovenous 0_2 content difference (vol%) and was divided by heart rate to give total VO_2 per beat (in $ml0_2$/beat). LV VO_2 was calculated by subtracting RV unloaded VO_2 from total VO_2, because the RV was kept mechanically unloaded and collapsed by continuous hydrostatic drainage. RV unloaded VO_2 was calculated as follows. Under baseline conditions, it was taken as biventricular VO_2 x the ratio of RV weight to biventricular weight. During BDM administration, it was taken as RV unloaded VO_2 under baseline conditions x the ratio of total VO_2 during BDM to total VO_2 under baseline conditions. For BDM data, this correction assumes that the drug decreases both LV and RV VO_2 proportionally. LV VO_2 was expressed as ml02/beat per 100g of LV.

Force time integral (FTI) is the time integral of total ventricular wall force through one cardiac cycle. Total wall force (F, in g) was calculated as 1.36 ($g \cdot cm^{-2} \cdot mmHg^{-1}$) x ventricular pressure (P, in mmHg) and lumen cross-sectional area (A, cm^2) based on the force equilibrium equation for a sphere. Because we used isovolumic contractions, we considered lumenal area to be constant throughout the cardiac cycle. Thus, $F=1.36 \times P \times A = 1.64 \times P \times V^{2/3}$, where V is volume.

Statistics

Data are presented as mean $\pm$ SD. $P < .05$ was considered statistically significant. The relation between VO_2 and FTI per beat at V_H and V_L under baseline conditions and at [BDM] < 5 mM was assessed by linear regression analysis. In individual hearts, repeated measures analysis of variance with multiple linear regression and a dummy variable was used to compare the slope and VO_2 intercept of the regression lines obtained at V_H and V_L. The regression model was

$$VO_2 = a_H + b_H \cdot FTI + a_\Delta \cdot D + b_\Delta \cdot D \cdot FTI \qquad (1)$$

where D is a dummy variable that encodes LV volume (i.e. D=0 if V_H, D=1 if V_L) and a_H, b_H, a_Δ, and b_Δ are regression coefficients. Coefficients a_H and b_H are the VO_2 intercept and slope, respectively, of the VO_2-FTI relation at V_H. Coefficient a_Δ is an estimate of the difference in the VO_2 intercept between V_H and V_L.

Coefficient b_Δ is an estimate of the difference in slope of the VO_2-FTI relation between V_H and V_L. A t test was used to examine whether the values of a_Δ and b_Δ were significantly different from zero. The t value was calculated as the estimate of the coefficient divided by the associated standard error. Comparisons of paired variables between V_H and V_L were performed by paired t test.

The fact that we obtained only five or six data points at each LV volume in each experiment limited our ability to detect differences between V_H and V_L in individual hearts. Therefore, to provide a more powerful test of whether any differences were present, all VO_2 and FTI data points were pooled and subjected to a repeated-measures analysis of variance with multiple linear regression. To eliminate inter-subject variability from the analysis, we defined 15 dummy variables and fitted the pooled data to another regression model

$$VO_2 = A_H + \sum_{i=1}^{15} a_i S_i + \sum_{i=1}^{15} b_i S_i \cdot FTI + B_H \cdot FTI + A_\Delta \cdot D + B_\Delta \cdot D \cdot FTI \qquad (2)$$

where S_i represents the dummy variables that account for between-subject variability [specifically, $S_i = 1$ if rabbit i, $S_i = -1$ if rabbit 16, $S_i = 0$ otherwise (for i=1-15)], and D is the same dummy variable as in Eq.1. Coefficients A_H and B_H are the overall VO_2-axis intercept and the overall slope, respectively, of the pooled VO_2-FTI relation for V_H. Coefficient A_Δ estimates the overall difference in VO_2-axis intercept between V_H and V_L, and coefficient B_Δ estimates the overall difference in slope of the VO_2-FTI relation between V_H and V_L. Our principal questions were whether A_Δ and B_Δ were significantly different from zero. We used a t test to assess differences in these coefficients.

Repeated-measures analysis of variance followed by the Student-Newman-Keuls test were applied to compare the unloaded VO_2 and the BDM estimate of nonmechanical VO_2 at V_H and V_L.

RESULTS

V_H ranged from 1.75 to 2.74 ml (avg 2.24 $\pm$ 0.32 ml). At V_H, LV end-diastolic pressure (LVEDP) averaged 9.7 $\pm$ 4.6 mmHg. V_L ranged from 0.95 to 1.75 ml (avg 1.48 $\pm$ 0.23 ml), at which LVEDP averaged 3.8 $\pm$ 2.8 mmHg. The average V_0 was 0.24 $\pm$ 0.09 ml. LV peak systolic pressure at V_L averaged 64.5 $\pm$ 6.7% of that at V_H.

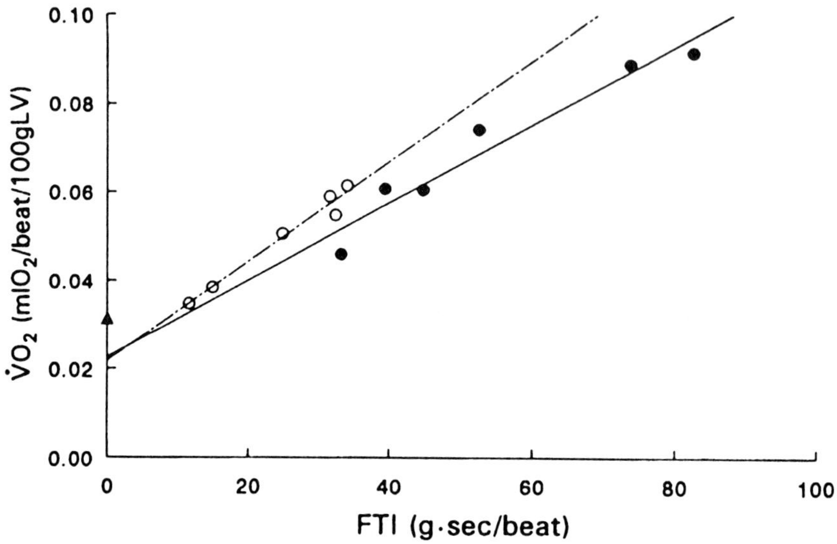

Fig. 1. Plot of representative O_2 consumption (VO_2)-force-time integral (FTI) relation and directly measured unloaded VO_2 (Δ) in one heart at left ventricular (LV) volume of 2.44 (●) and 1.44 ml (○). VO_2-FTI regression lines are shown for both volumes. Reprinted with permission from ref. 13.

The VO_2-FTI relations during BDM administration at V_H and V_L in a representative heart are shown in Fig. 1. In this case, V_H was 2.44 ml and V_L was 1.44 ml. VO_2 was closely correlated with FTI at both LV volumes, with the correlation coefficient for V_H being 0.970 and for V_L 0.985. The VO_2 axis intercept of the VO_2-FTI relation was 0.0226 and 0.0219 ml $O_2 \cdot beat^{-1} \cdot 100$ g LV^{-1} at V_H and V_L, respectively, and these values were not significantly different from each other ($P = 0.909$). The slope of the VO_2-FTI relation was 8.77×10^{-4} and 11.29×10^{-4} ml $O_2 \cdot g^{-1} \cdot s^{-1} \cdot 100$ g LV^{-1} for V_H and V_L, respectively. The difference in slope between V_H and V_L was not statistically significant ($P = 0.228$) in this case. The mechanically unloaded VO_2 (0.0313 ml $O_2 \cdot beat^{-1} \cdot 100$ g LV^{-1},) was higher than the BDM-derived VO_2-axis intercept value for V_H and V_L.

For all hearts, the mean correlation coefficient for the VO_2-FTI relation during BDM administration was 0.942 ± 0.031 at V_H and 0.926 ± 0.043 at V_L. The average VO_2-axis intercepts of the VO_2-FTI relation for V_H and V_L were virtually identical (0.0137 ± 0.0083 vs. 0.0132 ± 0.0090 ml $O_2 \cdot beat^{-1} \cdot 100$ g LV^{-1}, $P = 0.702$).

The slope of the VO_2-FTI relation for V_H and V_L was not significantly different in 14 of 16 hearts. However, the average slope value for V_L ($11.79 \pm 4.53 \times 10^{-4}$ ml $O_2 \cdot g^{-1} \cdot s^{-1} \cdot 100$ g LV^{-1}) was significantly greater than the value for V_H (8.23 ± 2.23 vs $11.79 \pm 4.53 \times 10^{-4}$ ml $O_2 \cdot g^{-1} \cdot s^{-1} \cdot 100$ g LV^{-1}, $P < 0.001$).

Pooled VO_2-FTI data from all 16 hearts are plotted in Fig. 2. Although there is considerable scatter of the data because of variability between subjects, VO_2 was significantly correlated with FTI at V_H and V_L ($r = 0.803$ for V_H and 0.718 for V_L). The results of the multiple linear regression model described in Eq. 2 are summarized in Table 1. The regression model enabled us to assess the overall slope and VO_2-axis intercept of the VO_2-FTI relation and eliminate between subject variability. A_Δ, an estimate of the difference in the VO_2-axis intercept between V_H and V_L, was not statistically significantly different from 0 ($P = 0.361$). In other words, the overall VO_2-axis intercept did not differ significantly between V_H and V_L. B_Δ, an estimate of the overall difference in slope of the VO_2-FTI relation between V_H and V_L, was significantly greater than zero ($P < 0.001$). Thus, the slope of the VO_2-FTI relation for V_L was significantly steeper than that for V_H.

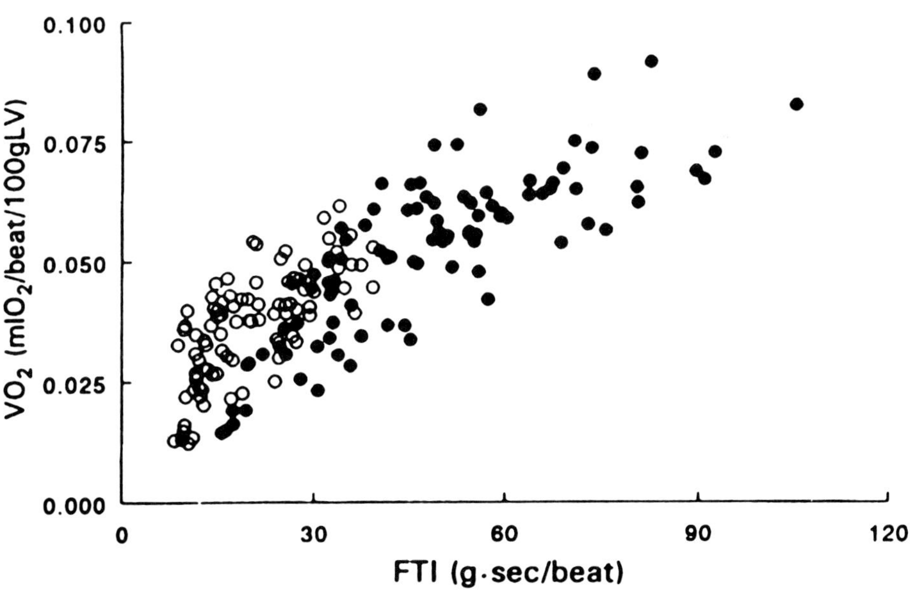

Fig. 2. Plot of pooled VO_2-FTI data at high volume (●) and low volume (○) from all 16 hearts. Reprinted with permission from ref. 13.

Table 1. Summary of multiple linear regression fits to Eq. 2.

r	*P*
$A_H = 0.0129 \pm 0.0015$ ml $0_2 \cdot beat^{-1} \cdot 100gLV^{-1}$	<0.001
$B_H = 8.274X10^{-4} \pm 0.315X10^{-4}$ ml $0_2 \cdot g^{-1} \cdot s^{-1}$ 100 g LV^{-1}	<0.001
$A_\Delta = 0.0017 \pm 0.0018$ ml $0_2 \cdot beat^{-1} \cdot 100gLV^{-1}$	0.361
$B_\Delta = 2.619X10^{-4} \pm 0.632X10^{-4}$ ml $0_2 \cdot g^{-1} \cdot s^{-1} \cdot 100gLV^{-1}$	<0.001

Values are means±SE.

To further assess the influence of LV volume on nonmechanical VO_2, we also determined whether there was a correlation between absolute values of LV volume and the corresponding VO_2-axis intercept of the VO_2-FTI relation obtained during BDM administration in all hearts. No significant correlation was found ($r = 0.094$, $n = 32$, $P = 0.611$).

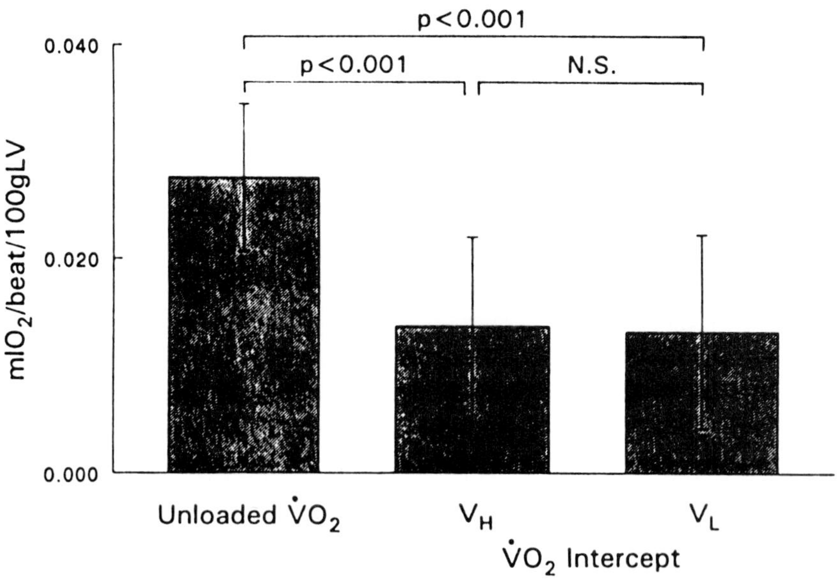

Fig. 3. Comparison of mechanically unloaded VO_2 and VO_2-axis intercept of VO_2-FTI relation (VO_2 intercept) at high (V_H) and low LV volume (V_L). Values are means ± SD. Reprinted with permission from ref. 13.

Figure 3 compares mean mechanically unloaded VO_2 and the mean VO_2-axis intercept of the VO_2-FTI relation at V_H and V_L. Mechanically unloaded VO_2 was about twice the value of the VO_2-axis intercept of the VO_2-FTI relation for V_H and V_L. Repeated-measures analysis of variance revealed this difference to be significant ($P < 0.001$). VO_2 for basal metabolism was 1.031 ± 0.542 and 1.020 ± 0.523 ml $O_2 \cdot min^{-1} \cdot 100$ g LV^{-1} for V_H and V_0, respectively. These values were not significantly different ($P = 0.735$).

DISCUSSION

The major finding of this study was that the VO_2-axis intercept of the VO_2-FTI relation obtained during BDM administration at concentrations less than 5 mM was virtually identical at V_H and V_L and, furthermore, the VO_2-axis intercept was not significantly correlated with LV volume. Using this value as an estimate of nonmechanical VO_2, i.e. VO_2 used for the combination of basal metabolism and EC coupling, we conclude that VO_2 for nonmechanical energy utilization is not importantly affected by LV volume in our preparation over a physiologic range of volumes. We also found that the slope of the VO_2-FTI relation obtained during BDM at V_L was significantly greater than at V_H and that the mechanically unloaded VO_2 was about twice as large as the BDM estimate of nonmechanical VO_2. With respect to the latter finding, we recently reported a similar observation (13), but the large number of hearts in this study allow us to estimate the difference between mechanically unloaded VO_2 and the BDM estimate with greater confidence. Assuming that nonmechanical VO_2 at V_0 is the same as that present at V_L and V_H, the difference between the mechanically unloaded VO_2 and the BDM estimate of nonmechanical VO_2 represents energy consumption for crossbridge cycling in the mechanically unloaded ventricle. Thus, the mechanically unloaded VO_2 appears to contain a substantial proportion of energy for residual crossbridge cycling. In this case, we mean substantial in relation to energy used for other nonmechanical purposes, i.e. EC coupling and basal metabolism. The proportion of energy consumed for crossbridge cycling under mechanically unloaded conditions remains relatively small in relation to total energy consumption under routine loading conditions in the isolated heart (for example, see Fig. 1).

The difference in slope of the VO_2-FTI relations obtained at V_H and V_L during BDM can most probably be ascribed to a difference in the way VO_2 is used for crossbridge cycling between mechanical unloading and BDM. To illustrate this point, consider two beats with the same FTI, one the non-BDM baseline beat at V_L, and the other a BDM beat at V_H (see Fig. 1). Compared with the baseline beat at V_H, FTI is lower in one case because BDM has been administered at constant volume, whereas in the other case the ventricle has been partly mechanically unloaded (from V_H to

V_L). Our results indicate that nonmechanical VO_2 is the same for both beats. Because FTI is the same, it seems reasonable to assume that the number of force-producing crossbridges is also the same. To account for the higher VO_2, we must postulate that a certain number of non-force-producing crossbridges are formed and split ATP under loaded conditions and that there are more of these for the baseline V_L beat. Because our results show that during mechanically unloaded conditions, there is considerable energy utilization for what we have termed "residual" crossbridge cycling, it makes sense that as the ventricle is loaded, there will continue to be a population of crossbridges that do not produce a net systolic force, although we have no way of quantifying this population under loaded conditions. The mechanism of absent or reduced net force production by some crossbridges could be related at least in part to spatial orientation, such that these crossbridges or myofilaments are aligned off the main fiber direction and/or their force production is wholly or partly cancelled by oppositely aligned crossbridges. Another possibility is that these crossbridges generate mechanical energy which is stored and released in the form of elastic recoil during diastole. Intuitively, a difference in non-force-producing crossbridges seems reasonable. One would not expect a drug that inhibits crossbridge formation to preferentially affect force- vs. non-force-producing crossbridges; i.e. as [BDM] is increased, both types of crossbridge should be equally inhibited. In contrast, a purely mechanical means of decreasing FTI (decreasing volume) might influence the spatial orientation of the myofibers, resulting in a different ratio of force- to non-force-producing crossbridges compared with BDM.

In a prior study using the BDM method (13) we quantified VO_2 for nonmechanical activity, but only at one ventricular volume. Using indo 1 as an indicator, we also documented that, at the concentrations employed in this study, BDM has no effect on the intracellular Ca^{2+} transient in the isolated rabbit heart. In most whole heart studies performed previously, VO_2 for nonmechanical energy utilization has been assessed by simply decreasing LV volume to a point at which peak systolic pressure becomes zero (i.e. mechanical unloading). However, Yasumura et al.(9) were able to produce unloaded contractions with different end-diastolic volumes in excised canine ventricles by rapidly withdrawing volume beginning very early in systole. Their results indicated that nonmechanical energy assessed in this fashion was no more than minimally affected by preload. However, regardless of how the LV is mechanically unloaded, shortening of myocardial fibers occurs and crossbridge formation clearly is not eliminated. As discussed previously, the VO_2 obtained under such conditions must contain an uncertain amount used for crossbridge cycling and, therefore, overestimates VO_2 for nonmechanical energy utilization. In contrast, BDM at relatively low concentrations suppresses crossbridge cycling without importantly affecting Ca^{++} cycling in rabbit and human heart muscle. This is the principal underlying our use of the extrapolation of the VO_2-FTI relation obtained at low [BDM] to the zero FTI value as an estimate of VO_2 used for EC coupling and basal metabolism. Because we also showed that basal metabolic VO_2 did not vary with LV volume, we conclude that under the conditions of this study, VO_2 used for EC coupling is not importantly influenced by LV volume.

In heart muscle, BDM inhibits EC coupling in addition to crossbridge cycling. In rabbit, however, these effects become energetically significant only at concentrations > 5-6 mM (8). In other species, however, EC coupling inhibition may occur to a significant extent at lower concentrations (8,15-17). Although we (13) and Alpert et al. (8) have carefully validated the BDM method in rabbit heart and shown that it provides a more reasonable estimate of activation energy than mechanical unloading, we cannot completely exclude the possibility of some effects on EC coupling at the low concentrations which we have employed. Thus, while the evidence indicates that these effects are small at worst, to the extent that they occur and reduce Ca^{++} reuptake by the sarcoplasmic reticulum, this would result in depression of the VO_2-FTI relation during BDM administration and an underestimate of nonmechanical VO_2.

In isolated papillary muscle, previous reports have indicated that the intracellular Ca^{++} transient and energy use for EC coupling are dependent on muscle length (1-8). For example, Alpert et al. (3) reported a 22 $\pm$ 4% decrease in tension-independent heat of rabbit papillary muscle using the BDM method when muscle length was shortened by 11%. In our own study, the difference in ventricular volume from the low to high volume corresponded to a 12.9% change in circumference at the subendocardial layer, assuming that the LV can be regarded as a sphere. This apparent discrepancy between our results and those obtained in isolated muscle could be explained by a number of different experimental conditions, for instance temperature and contraction frequency. In addition, we did not specifically attempt to achieve the whole heart volume equivalent of L_{max}, as was the case in the isolated muscle studies. Additionally, the whole ventricle, in contrast to isolated muscle preparations, may have much greater spatial inhomogeneity of end-diastolic fiber length (18). Thus, our results may have been obtained over different ranges of average muscle length in relation to L_{max} than those employed in isolated muscle.

With respect to the significance of our results in relation to length-dependent activation, two mechanisms have been proposed for this phenomenon. One is an increase in Ca^{++} sensitivity of the myofilaments as muscle length increases (5,6). This mechanism has no direct energetic cost and is therefore consistent with our observations in the whole heart. The second proposed mechanism is an increase in systolic Ca^{++} release from the sarcoplasmic reticulum (1,3,7). This mechanism is supported by studies in which the intracellular Ca^{2+} transient has been observed to increase following an increase in muscle length (3). Increased Ca^{++} release should in turn result in a greater amount of Ca^{++} reuptake. Assuming that most reuptake is related to the activity of the sarcoplasmic reticulum Ca^{++} ATPase such a mechanism would directly increase the energy costs of EC coupling. Correspondingly, the aforementioned length-dependence of tension-independent heat is consistent with this particular mechanism of length-dependent activation (8). Since we found that energy for EC coupling did not increase with increased volume, this suggests that in the whole heart increased Ca^{++} cycled per beat is not an important mechanism of length-dependent activation. A possible explanation for this isolated muscle-whole heart discrepancy is that the balance between the two mechanisms is altered in relation to the aforementioned differences in experimental conditions which are present in isolated muscle and whole heart studies.

Suga et al. (11) have proposed the use of PVA as an expression of the total mechanical energy of contraction (external work + potential energy). The efficiency of the contractile machinery using this concept is the ratio of PVA to mechanical VO_2 or the inverse slope of the linear VO_2-PVA relation. In our studies (10,19) and those of Suga (11), mechanical unloading has been used to partition VO_2 and in turn to calculate contractile efficiency using the VO_2-PVA framework. As has been discussed previously, the unloaded VO_2 contains significant crossbridge cycling energy and overestimates the true nonmechanical VO_2. However, our results do not necessarily mean that the VO_2-PVA relation obtained using mechanical unloading can no longer be used to estimate changes in contractile efficiency or nonmechanical VO_2. This point is illustrated in Fig. 4. In this figure we show that the mechanically unloaded VO_2 overestimates VO_2 used for nonmechanical purposes and therefore results in an underestimation of VO_2 used for crossbridge cycling. Within the VO_2-PVA framework, contractile efficiency is the total energy output of the crossbridges (PVA) divided by their energy input (mechanical VO_2). Therefore, mechanical unloading results in an overestimate of efficiency. Because we found that nonmechanical VO_2 is not importantly influenced by diastolic ventricular volume, the slope of the VO_2-PVA relation generated by varying ventricular volume should still reflect changes in contractile efficiency, providing the intervention studied does not itself result in a significant dependence of nonmechanical VO_2 on ventricular volume. Similarly, the mechanically unloaded VO_2 should still reflect changes in nonmechanical VO_2, providing the intervention studied does not influence the amount of residual crossbridge cycling at V_0.

Finally, we would propose a modification of the VO_2-PVA approach based on the concept enunciated previously. As the ventricle is loaded, the mechanical VO_2 should be considered as containing energy used by both force- and non-force-producing crossbridges. At present, we have no way to quantify the ratio of the two. However, it would seem likely that the VO_2 used by non-force-producing crossbridges should be a linear function of PVA. This is because the total VO_2-PVA

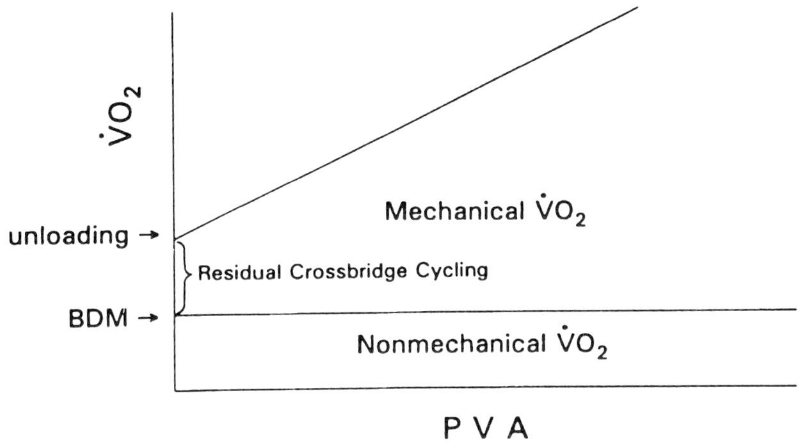

Fig. 4. Schematic diagram of relation between VO_2 per beat and pressure-volume-area (PVA) relation. Solid diagonal line, VO_2-PVA line obtained by mechanical unloading of ventricle. Mechanically unloaded VO_2 (unloading) is subdivided into VO_2 for residual crossbridge cycling and that for nonmechanical activity (nonmechanical (VO_2). On the basis of our results, nonmechanical VO_2 obtained by 2,3-butanedione monoxime (BDM) method, which consists of VO_2 for excitation-contraction coupling and basal metabolism, is constant regardless of PVA. Reprinted with permission from ref. 13.

relation is linear; hence, if non-force-producing crossbridge VO_2 increased non-linearly, this would result in the unlikely coincidence of two nonlinear relationships summating to a linear relationship.

ACKNOWLEDGEMENTS

We thank Stephen P. Bell and Judit Fabian for excellent technical assistance and Dr. Bryan K. Slinker for help in statistical analysis.

This study was supported by National Heart, Lung, and Blood Institute Grant HL-45116 and National American Heart Association Grant #93006340.

REFERENCES

1. Fabiato, A., and Fabiato, F. Dependence of the contractile activation of skinned cardiac cells on the sarcomere length. Nature Lond. 256:54-56,1975.

2. Cooper, G. Myocardial energetics during isometric twitch contractions of cat papillary muscle. Am J Physiol 236 (Heart Circ Physiol 5):H244-H253, 1979.

3. Allen, D.G., and Kurihara, S. The effects of muscle length on intracellular calcium transients in mammalian cardiac muscle. J Physiol Lond 327:79-94,1981.

4. Cooper, G. Influence of length changes on myocardial metabolism in the cat papillary muscle. Circ Res 49:423-433, 1981.

5. Hibberd, M.G., and Jewell, B.R.. Calcium- and length-dependent force production in rat ventricular muscle. J Physiol Lond 329:527-540,1982.

6. Blinks, J.R., and Endoh, M. Modification of myofibrillar responsiveness to Ca^{++} as an inotropic mechanism. Circulation 73, Suppl III:III-85-III-98, 1986.

7. Allen, D.G., Nichols, C.G., and Smith. G.L. The effects of changes in muscle length during diastole on the calcium transient in ferret ventricular muscle. J Physiol Lond 406:359-370, 1988.

8. Alpert, N.R., Blanchard, E.M., and Mulieri, L.A. Tension-independent heat in rabbit papillary muscle. J Physiol Lond 414:433-453, 1989.

9. Yasumura, Y., Nozawa, T., Futaki, S., Tanaka, N. and Suga, H. Minor preload dependence of O_2 consumption of unloaded contraction in dog heart. Am J Physiol 256 (Heart Circ Physiol 25):H1289-H1294, 1989.

10. Goto, Y., Slinker, B.K., and LeWinter, M.M. Similar normalized E_{max} and O_2 consumption-pressure-volume-area relation in rabbit and dog. Am J Physiol 255 (Heart Circ Physiol 24):H366-H374, 1988.

11. Suga, H. Ventricular energetics. Physiol Rev 70:247-277, 1990.

12. Nozawa, T., Yasumura, Y., Futaki, S., Tanaka, N. and Suga, H. No significant increase in O_2 consumption of KCl-arrested dog heart with filling and dobutamine. Am J Physiol 255 (Heart Circ Physiol 24):H807-H812, 1988.

13. Yaku, H., Slinker, B.K., Mochizuki, T., Lorell, B.H. and LeWinter, M.M. Use of 2,3-butanedione monoxime to estimate nonmechanical VO_2 in rabbit hearts. Am J Physiol 265 (Heart Circ Physiol 34):H834-H842, 1993.

14. Yaku, H., Slinker, B.K., Myhre, E.S.P., Watkins, M.W., and LeWinter, M.M. Stability of myocardial O_2 consumption-pressure-volume-area relation in red cell-perfused rabbit heart. Am J Physiol 261 (Heart Circ Physiol 30):H1630-H1635, 1991.

15. Blanchard, E.M., Smith, G.L., Allen, D.G., Alpert, N.R. The effects of 2,3-butanedione monoxime on initial heat, tension and aequorin light output of ferret papillary muscles. Pfluegers Arch 416:219-221, 1990.

16. Gwathmey, J.K., Hajjar, R.J. and Solaro, R.J. Contractile deactivation and uncoupling of crossbridges. Effects of 2,3-butanedione monoxime on mammalian myocardium. Circ Res 69:1280-1292, 1991.

17. Perreault, C.L., Mulieri, L.A., Alpert, N.R., Ransil, B.J., Allen, P.D.and Morgan, J.P. Cellular basis of 2,3-butanedione monoxime in human myocardium. Am J Physiol 263 (Heart Circ Physiol 32):H503-H510, 1992.

18. Yoran, C., Covell, J.W., and Ross, J.,Jr. Structural basis for the ascending limb of left ventricular function. Circ Res 32:297-303, 1973.

19. Goto, Y., Slinker, B.K. and LeWinter, M.M. Decreased contractile efficiency and increased nonmechanical energy cost in hyperthyroid rabbit heart. Relation between O_2 consumption and systolic pressure-volume-area or force-time integral. Circ Res 66:999-1011, 1991.

11. Emax and Myocardial Microcirculation

Fumihiko Kajiya, Toyotaka Yada, Yasuo Ogasawara, Shigeru Ohta, Osamu Hiramatsu, Masami Goto, Katsuhiko Tsujioka

INTRODUCTION

Cardiac contraction is one of the major determinants of instantaneous coronary blood flow. In 1695, Scaramucci first speculated that cardiac contraction squeezed myocardial vessels and caused the pulsation of coronary arterial blood flow (1). This hypothesis was later confirmed using newly-developed measuring tools that were applicable to the coronary blood flow of a beating heart (2,3,4). Modern ultrasound and laser technology enabled us to investigate the phasic pattern of the coronary inflow to the myocardium at the peripheral portion of the artery or in the septal artery (5,6); thus, the effects of cardiac contraction on myocardial perfusion can be evaluated without the influence of the compliance of extramural coronary arteries.

Figure 1 shows an example of velocity waveforms in a small epicardial artery (150-400 μm) of the left ventricle measured at a location just before its penetration into the myocardium, and in a small vein measured just after its emergence from the myocardium. These velocities were measured by a laser Doppler method with an optical fiber. The fiber tip (velocity sensor; core diameter 0.05 mm) was fixed on the outer surface of the vessel by a drop of cyanoacrylate. It is clearly shown that the phasic flows in the left coronary artery and vein are unlike those of other organs; the arterial flow is restricted to diastole, whereas the venous flow is restricted to systole.

To directly analyze the mechanism of coronary flow dynamics, it is necessary to observe subendocardial (deep myocardial) microcirculation in a beating heart. However, in vivo observation of the subendocardial circulation has been hampered mainly due to a lack of appropriate optical technology to investigate subendocardial vessels. Only the subepimyocardial microcirculation has been analyzed by a sophisticated microscope system. To overcome this limitation, we developed a portable needle-probe videomicroscope with a charge-coupled-device (CCD) camera and succeeded in obtaining access to subendocardial microvessels (7).

In this paper, following our description of the subendocardial microvessels, we evaluate diameter changes of the subendocardial microvessels during a cardiac cycle by a model experiment in relation to *cardiac contractility* (Emax) and *intramyocardial pressure pump* produced by left ventricular pressure.

METHODS

Observation of Subendocardial Microvessels

Observation of the subendocardial arteriole of a beating porcine heart was carried out using the needle-probe videomicroscope, previously described in detail (8). In brief, the system consists of a needle-probe, a camera body containing a CCD camera, a lens and light guides, a control unit, a light source, a monitor and a videocassette recorder (Fig. 2). The camera body has a 1/2-inch CCD image sensor, about 250,000 pixels and 330-line horizontal resolution . The needle-probe (diameter: 4.5 mm) contains a gradient index (GRIN) lens surrounded by an annular light guide. One end of the needle-probe is attached to the camera body with a focusing system. The image passes through the GRIN lens and focuses on the CCD image sensor.

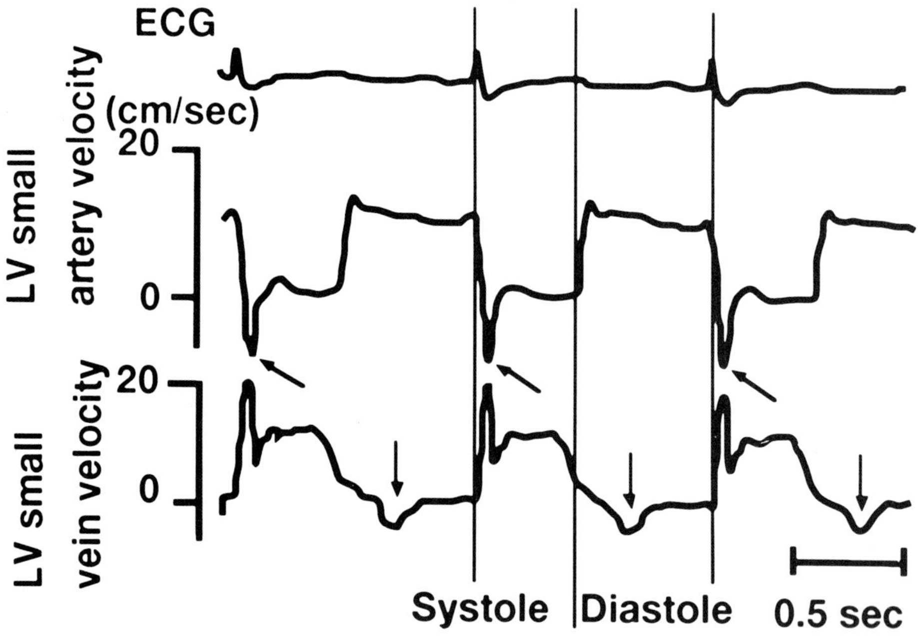

Fig. 1. An example of velocity waveforms in a small epicardial artery of the left ventricle just before its penetration into the myocardium, and in a small vein just after its emergence from the myocardium. The velocity waveform of coronary arterial inflow is almost exclusively diastolic, while the coronary venous outflow is predominantly systolic. In addition, there were small reverse flows both in the arterial (systolic) and venous (diastolic) flows (arrow heads). Modified from Kajiya et al ref. 4.

The image is monitored and recorded on a videotape. The tissue is illuminated by light from a halogen lamp, which is transmitted through light guides surrounding the GRIN lens. A green filter is used to accentuate the contrast between the image of the vessel and the surrounding tissue. The needle-probe is enclosed in a silastic 14F double lumen sheath with a doughnut-shaped balloon at its tip (See Fig. 2).

Twelve pigs were anesthetized with ketamine and sodium pentobarbital. The right carotid artery and the right jugular vein were catheterized for hemodynamic and arterial-blood-gas measurements and for fluid and drug administrations. Blood pressure was measured in the ascending aorta and the left ventricle. Electrocardiogram was recorded by the standard leads. Following a median sternotomy and a left thoracotomy through the fifth intercostal space, the heart was exposed and suspended in a pericardial cradle. The needle-probe was introduced into the left

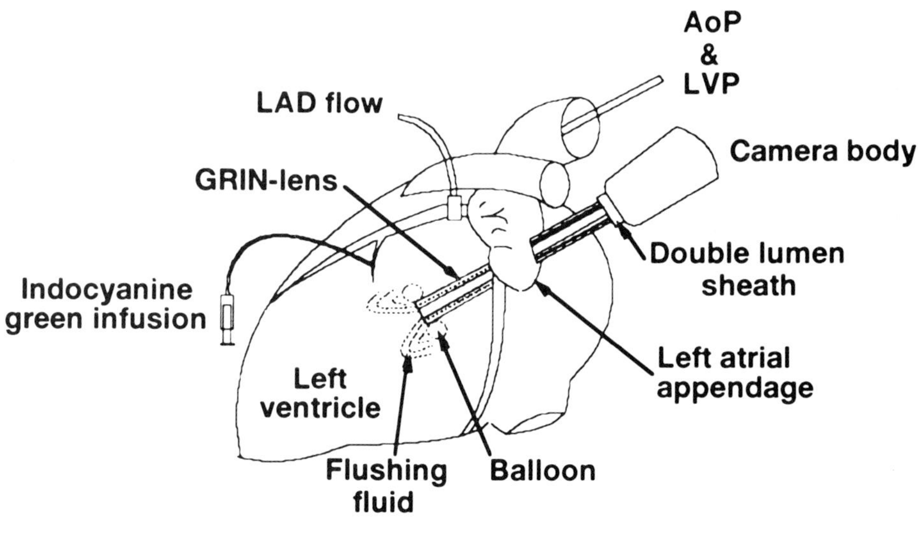

Fig. 2. Illustration of the needle-probe videomicroscope with a charge-coupled device (CCD) camera. The microscope system consists of a needle probe, a camera body containing a CCD camera, a lens and light guide, a control unit, a light source, a monitor, and videocassette recorder (VCR). A needle-probe containing a gradient-index (GRIN) lens with a length of 180 mm is used to obtain the images of the subendocardial microcirculation of the left ventricle. Reprinted with permission from ref. 8.

ventricle through an incision in the left atrial appendage via the mitral valve (Fig. 3). The inflated doughnut balloon was gently placed against the endocardial surface and the intervening blood was flushed away with a Krebs-Henseleit buffer solution through a microtube to obtain a clear image.

Model Analysis

We adopted a two vascular-layer model (Fig.4), in which the epimyocardial vascular layer was located at 1/4 depth from the epicardium and the endomyocardial layer at 1/4 depth from the endocardium (9). Each vascular layer was composed of arterioles, capillaries and venules, which were characterized by a modified Windkessel model with volume-dependent resistance and capacitance as proposed by Bruinsma et al. (10).

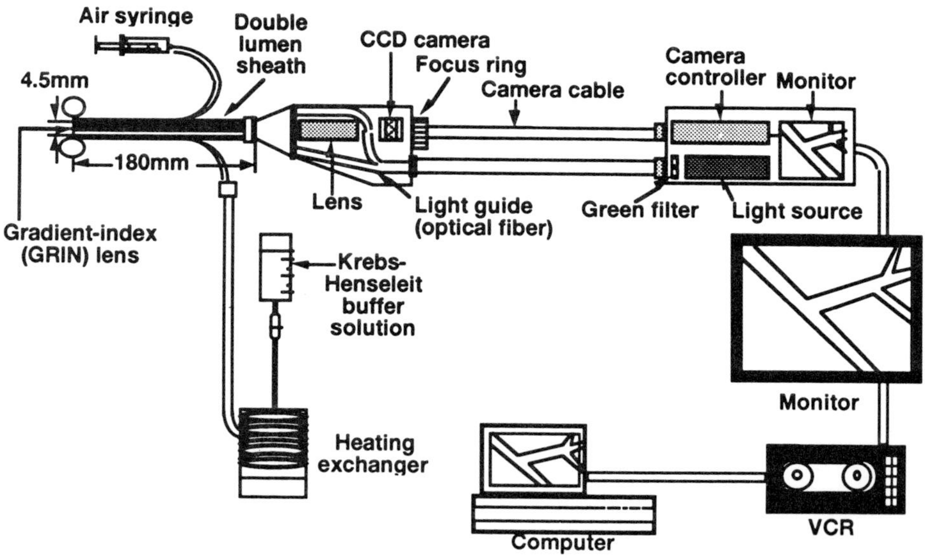

Fig. 3. Experimental preparation to obtain access of the needle-probe to subendocardial arterioles and venules. The sheathed needle-probe was introduced into the left ventricle through an incision in the left atrial appendage via the mitral valve. The balloon was inflated to prevent direct compression of the endocardial microcirculation by the needle tip, and the intervening blood between the needle-probe and endocardium was flushed away with a Krebs-Henseleit buffer solution. Reprinted with permission from ref. 8.

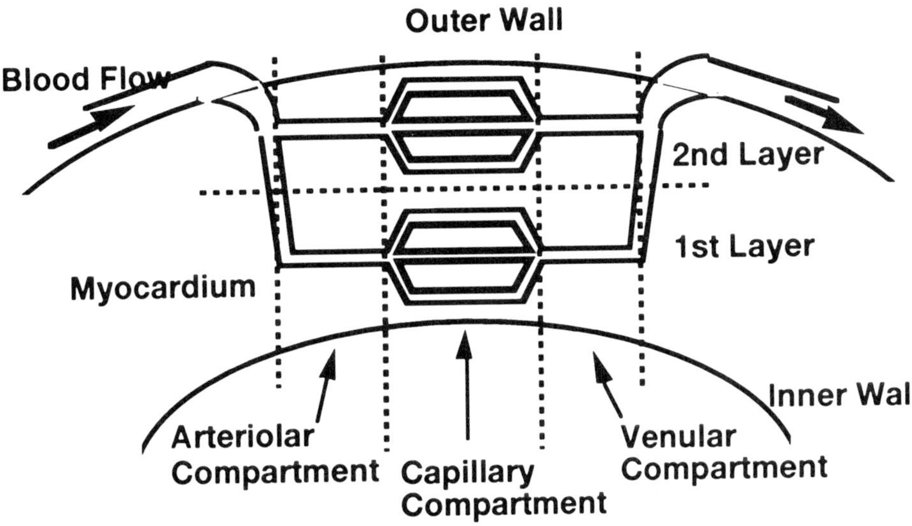

Fig. 4. Three compartment, two layer model to simulate the intramyocardial vascular network. The epimyocardial vascular layer was located at 1/4 myocardial depth from the epicardium and the endomyocardial layer at 1/4 depth from the endocardium. Reprinted with permission from ref. 9.

To quantitate the intramyocardial vascular compressive force, we combined the intramyocardial pump model of Spaan et al. (11,12) with the time-varying elastance model originally proposed by Suga (13,14) (Fig. 5). The change in myocardial elastance was expressed by altering the relation between vascular volume and transmural pressure (P_{TR}). We assumed that 75% and 25% of left ventricular (LV) pressure were applied respectively to endomyocardial and epimyocardial vessels as intramyocardial pressure (P_{IM}). Aortic and LV pressures were used as the input to the model.

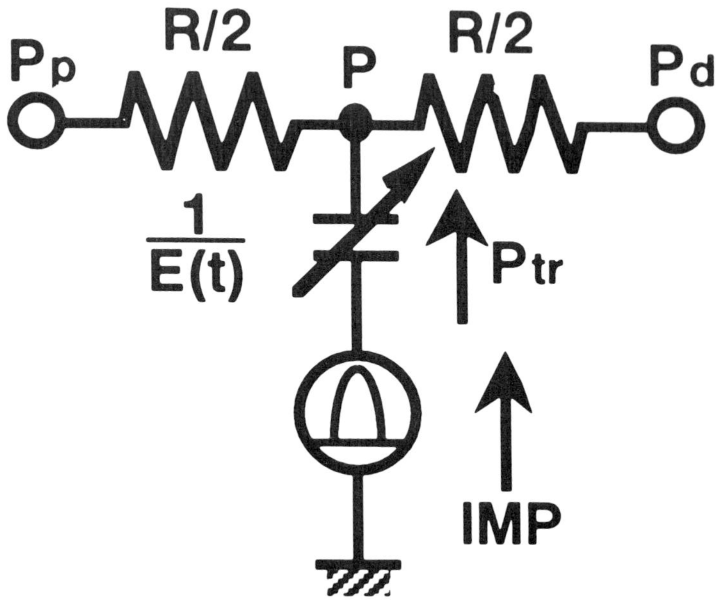

Fig. 5. Electrical analog model of each vascular compartment. We assumed that 75% and 25% of the left ventricular pressure is applied to the endomyocardial and epimyocardial vessels as intramyocardial pressure (IMP). Ptr= transmural pressure, Pp= proximal pressure, Pd= distal pressure, P= pressure in each vascular compartment, R= resistance. Reprinted with permission from ref. 9.

RESULTS

Images of Subendocardial Arterioles and Venules

We identified subendocardial microvessels in all cases studied, although it was difficult to find an appropriate vessel in some cases due to wide variation in vessel distribution in the subendocardium. Figure 6 shows typical images of a subendocardial arteriole and venule at end-systole and end-diastole. Arterioles and venules were differentiated by an injection of indocyanine green. Following a dye injection, the image of arterioles appeared before venules. Although the diameters of subendocardial arterioles and venules decreased significantly from end-diastole to end-systole as can be seen in Fig. 6, we usually did not observe any visible collapse of these vessels during systole and also during a prolonged diastole (15). The image density of both arterioles and venules decreased at end-systole in comparison with that at end-diastole in all cases. This may indicate that the diameter in the vertical direction also decreases from end-diastole to end-systole.

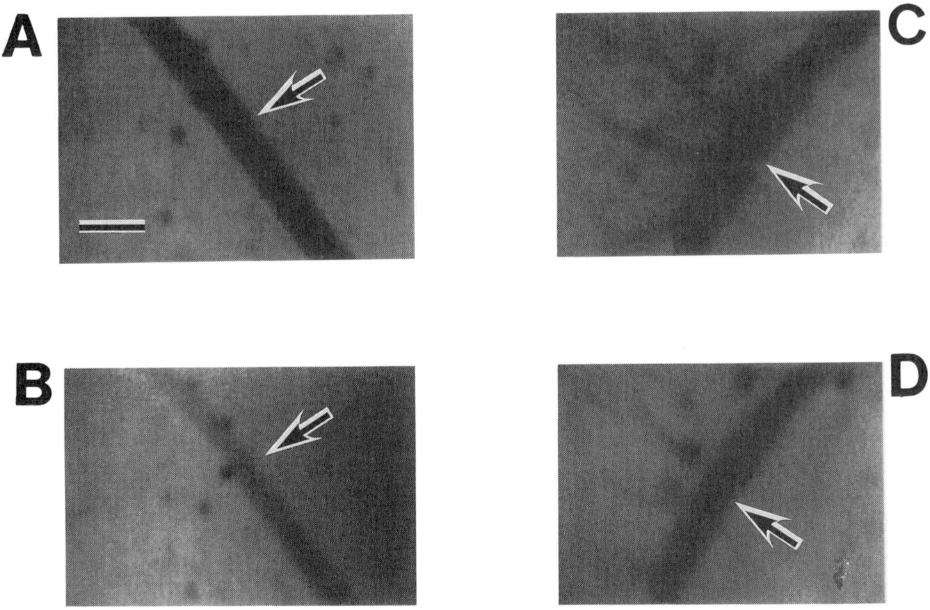

Fig. 6. Images of a subendocardial arteriole (left) and venule (right) at end-diastole and end-systole. A, image of a subendocardial arteriole at end-diastole, arrowhead = 120 μm; B, image of a subendocardial arteriole at end-systole, arrowhead = 90 μm; C, image of a subendocardial venule at end-diastole, arrowhead = 195 μm; D, image of a subendocardial venule at end-systole, arrowhead = 135 μm.

Figure 7 shows the diameter changes during a cardiac cycle in arterioles and venules plotted against their end-diastolic diameters. The diameter was measured for vascular segments with a relatively uniform change along the vessel. On an average, the arteriolar diameter decreased by about 24%, while the venular diameter decreased by about 17%. The vessels with smaller diameters were less pulsatile than larger vessels in both arterioles and venules.

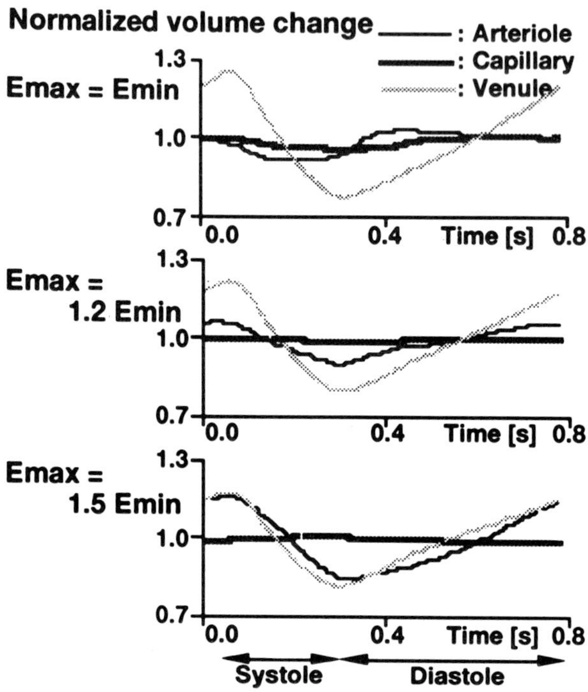

Fig. 7. Correlation between diameter at end-diastole (horizontal axis) and corresponding percent change in diameter from end-diastole to end-systole (vertical axis). Panel A: Subendocardial arterioles; panel B: subendocardial venule. The percentage change in the diameter showed a significant correlation with the end-diastolic vascular diameter both in subendocardial arterioles and venules. The horizontal dotted lines indicate the mean values of diameter changes. Reprinted with permission from ref. 8.

Model Interpretation

Figure 8 shows the simulation results of arteriolar, capillary and venular volume changes in three different states of time-varying elastance, i.e., (i) the time-varying elastance was kept constant at its end-diastolic value during a cardiac cycle (upper panel), (ii) Emax at end-systole increased to a value 1.2 times greater than the end-diastolic elastance (middle panel), (iii) Emax increased to a value 1.5 times greater than the end-diastolic elastance (lower panel). It should be noted that the pulsation amplitude of venules was much greater than that of arterioles when the elastance was kept at a low, constant value throughout a cardiac cycle, while the pulsation amplitude of arterioles and venules was similar when Emax was set at a value 1.5 times higher than the end-diastolic elastance. Thus, introduction of Emax as the extravascular compressive force in addition to intramyocardial pressure produced by

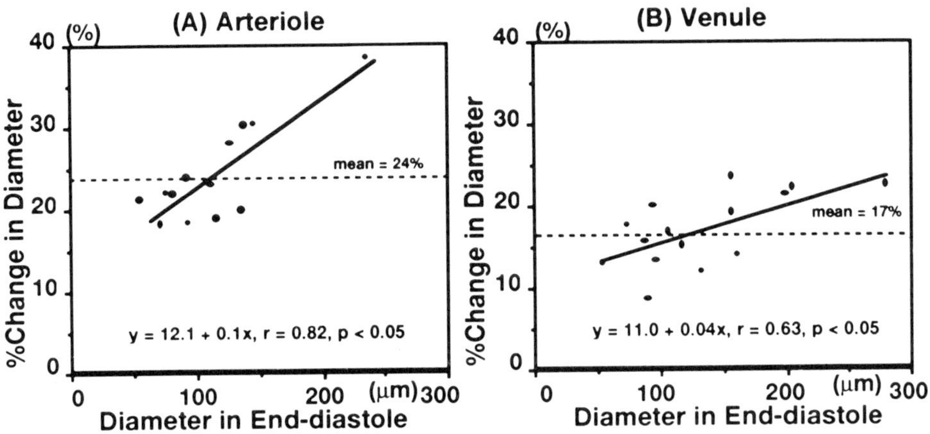

Fig. 8. Simulation results of vascular volume in subendomyocardial arteriole, capillary and venule. Upper panel: Myocardial elastance was kept constant at a low level equal to the minimum end-diastolic value (Emin) through a cardiac cycle, middle panel: Emax at end-systole was set at a value 1.2 times Emin, lower panel: Emax = 1.5 times Emin.

LV pressure is necessary to explain the similar change in arteriolar and venular diameters (both 20%). The diameter change in the capillary by the model prediction with any Emax value was very small, which was consistent with our experimental results, i.e., the smaller diameter, the less pulsatile (see Fig. 7). The diastolic predominance of coronary arterial flow could be simulated with any Emax value.

DISCUSSION

The subendocardial microvessels were successfully visualized in the beating heart. Arteriolar and venular diameter changes during a cardiac cycle were about 20%, which may explain the diastolic-predominant arterial flow and the systolic-predominant venous flow.

These changes in subendocardial arterioles and venules differ greatly from those of the epimyocardial microcirculation. Kanatsuka et al.(16) observed only a 1.1% diameter change of the left ventricular epimyocardial arterioles during a cardiac cycle as measured by an intravital microscopic system. Nellis et al.(17) reported a 20% increase in right ventricular venular diameter just after end-systole as measured by a transillumination microscope with a cardiac cycle-synchronized light source. Using systolic and diastolic arrested hearts, Goto et al.(18) showed that cardiac contraction affected deep myocardial small arterioles and capillaries predominantly. Judd and Levy (19) used arrested hearts of several species under various experimental conditions, and found a significant decrease in vascular volume by cardiac contraction. The results of both Goto et al. and Judd and Levy are consistent in showing a significantly smaller diameter of subendocardial vessels with systolic arrest than with diastolic arrest, and they are consistent with our present findings.

The larger subendocardial arterioles and venules show more phasic diameter changes than the smaller ones. Judd and Levy (19) also observed a greater vascular volume change between systolic and diastolic arrested rat hearts in vessels with diameters larger than 100 μm compared with smaller vessels ($>$100 μm). Our observation is consistent with theirs. The main reason for this observation may be resistance in smaller vessels, since the blood contained in high resistance vessels is harder to displace by squeezing than blood in low resistance vessels. The high resistance of smaller vessels also implies that they may have a long time constant (resistance times compliance), assuming a similar compliance value between large and small vessels. Long time constants in small vessels will result in less phasic diameter change.

The blood volume change between diastole and systole in the myocardium has been explained by two models: the intramyocardial pressure pump model of Spaan et al. (11,12) and the time-varying elastance model of Krams et. al. (13) based on a model originally proposed by Suga (14). According to the intramyocardial pressure pump model, with a constant, minimum elastance in which 75% and 25% of LV pressure is applied as the intravascular compressive force to the subendomyocardial and subepicardial vessels, compression of the venular compartment during systole is much more vigorous than that by the time-varying elastance model (the upper panel of Fig. 8). On the other hand, addition of the time-varying elastance with Emax of 1.5 times greater than the end-diastolic elastance value could explain the similar degree of diameter change between arterioles and venules (the lower panel of Fig. 8). According to the elastance model, the behavior of the subepicardial vessels should be similar to that of the subendocardial vessels; however, this is not the case. As mentioned above, the diameter of subepicardial arterioles is almost unchanged between end-diastole and end-systole, unlike the subendocardial arterioles.

This may be mainly due to the less powerful, systolic extravascular compressive force of epimyocardium associated with left ventricular pressure, which may interact with an elastance effect. As for the venules, the subepicardial venular diameter increased by about 20%, while the subendocardial venules decreased by about 20%. This may be explained as follows: the systolic increase in venous outflow from deeper myocardial layers to the epicardial veins and some resistance within the epicardial veins may result in an increase in intraluminal venous pressure and hence an increase in epicardial and subepicardial venular diameters during systole.

In conclusion, vascular compression by cardiac contraction decreased the diameter of subendocardial arterioles and venules by about 20% with a tendency for a smaller percent diameter change in small vessels. These diameter changes explain the phase-difference between diastolic coronary arterial inflow and systolic venous outflow. Introduction of Emax was necessary to simulate the similar changes in subendocardial arteriolar and venular diameters.

ACKNOWLEDGEMENT

We are grateful to Dr. Takuji Okamoto, Department of Information Technology, Okayama University for his collaboration in the model experiment. We thank Dr. Seiichi Mochizuki for his useful comments and English revision, and Ms. Chikako Tokuda for her excellent technical and secretarial work. This study was supported by Research Project Grant from Kawasaki Medical School and by Grant-in-Aid 3-106 and 4-107 from the Ministry of Education, Science and Culture, Japan.

REFERENCES

1. Scaramucci J. Theoremate familiaria viros eruditos consulentia de variis physico-medicis lucubratiionibus juxta leges mecanicas. Apud Joannem Baptistam Bustum 1695:70-81.

2. Gregg D.E. The coronary circulation in health and disease. Lea & Febiger, Philadelphia, 1950.

3. Marcus ML. The coronary circulation in health and disease. McGraw-Hill, 1983.

4. Kajiya F., Klassen G.A., Spaan J.A.E., Hoffman J.I.E. Coronary circulation: basic mechanism and clinical relevance. Springer-Verlag, Tokyo, 1990.

5. Chilian W. and Marcus E.S. Phasic coronary flow by a vascular waterfall mechanism. Circ Res 50:775-781, 1982.

6. Kajiya F., Tomonaga G., Tsujioka K., Ogasawara Y., Nishihara H. Evaluation of local blood flow velocity in proximal and distal coronary arteries by laser Doppler method. J Biomech Eng 107:10-15, 1985.

7. Kajiya F., Goto M., Yada T., Kimura A., Yamamoto T, Hiramatsu O., Ogasawara Y., Tsujioka K., Yamamori S., Hosaka H. In vivo evaluation of endocardial blood vessels by a new needle type CCD microscope (abstract). Circulation 84:II-271, 1991.

8. Yada T., Hiramatsu O., Kimura A., Goto M., Ogasawara Y., Tsujioka K., Yamamori S., Ohno K., Hosaka H., Kajiya F. In vivo observation of subendocardial microvessels of the beating porcine heart using a needle-probe videomicroscope with a CCD camera. Circ Res 72:939-946, 1993.

9. Ohta S., Okamoto T., Ogasawara Y., Matsumoto T., Kajiya F. Simulation study of intramyocardial blood flow dynamics (in Japanese). Trans IECE Jap J77-D-II:441-448, 1994.

10. Bruinsma P., Arts T., Dankelman J., Spaan J.A.E. Model of the coronary circulation based on pressure dependence of coronary resistance and compliance. Basic Res Cardiol 83:510-52, 1988.

11. Spaan J.A.E., Breuls N.P.W., Laird J.D. Diastolic-systolic flow differences are caused by intramyocardial pump action in the anesthetized dog. Circ Res 49:584-593, 1981.

12. Spaan J.A.E. Coronary blood flow. Kluwer Academic Pub, Dordrecht, 1991.

13. Krama R., Sipkema P., Westerhof N. Can coronary systolic-diastolic flow difference be predicted by left ventricular pressure or time varying intramyocardial elastance? Basic Res Cardiol 84:149-159, 1989.

14. Suga H., Sagawa K. Instantaneous pressure-volume relationships and their ratio in the excised, supported canine left ventricle. Circ Res 35:117-126, 1974.

15. Hiramatsu O., Goto M., Yada T., Kimura A., Tachibana H., Ogasawara Y., Tsujioka K., Kajiya F. Diameters of subendocardial aterioles and venules during prolonged diastoles in canine left ventricles. Circ Res (in press).

16. Kanatsuka H., Lamping K.G., Eastham C.L., Dellsperger K.C., Marcus M.L. Comparison of the effects of increased myocardial oxygen consumption and adenosine on the coronary microvascular resistance. Circ Res 65:1296-1305, 1989.

17. Nellis H.N., Whitesell L. Phasic pressures and diameters in small epicardial veins of the unrestrained heart. Am J Physiol 257:H1056-H1061, 1989.

18. Goto M., Flynn A.E., Doucette J.W., Jansen C.M.A., Stork M.M., Coggins D.L., Muehrcke D.D., Husseini W.K., Hoffman J.I.E. Cardiac contraction affects deep myocardial vessels predominantly. Am J Physiol 261:H1417-H1429, 1991.

19. Judd R.M., Levy B.I. Effects of barium-induced cardiac contraction on large and small-vessel intramyocardial blood volume. Circ Res 68:217-225, 1991.

12. Efficiency of Ventricular-Arterial Coupling and Baroreflex Regulation of Blood Pressure

Kenji Sunagawa, Masaru Sugimachi, Yasuhiro Ikeda,
Osamu Kawaguchi, Toshiaki Shishido and Toru Kawada

INTRODUCTION

The heart contracts incessantly over the lifetime. The average number of contractions for human beings exceeds 2.5 billion. Since each ejection requires a large amount of energy, the energy expenditure by the heart becomes enormous. Thus conceivable that, in response to changes in demands of the heart, the regulatory system adjusts ventricular contraction to maximize energy efficiency. Indeed, stroke power output of the normal excised feline left or right ventricle was maximum when the ventricle was loaded with normal arterial impedance (1-3). This was true in the open-chest cat (4-6). In the canine left ventricle, external work was expected to be nearly maximum under normal loading conditions (7,8), as was indeed the case (9-12). Although all these studies indicated that external work or mechanical efficiency was well optimized in conscious animals as well as in anesthetized animals, it was not known whether the optimization principle holds in animals under exercise stress, where the metabolic demand of the heart is greatly increased. Thus in chronically instrumented dogs, we investigated the effect of exercise on mechanical energy transmission from the left ventricle to the arterial system (13).

EFFICIENCY OF VENTRICULAR-ARTERIAL COUPLING IN CONSCIOUS DOGS

Methods

Six adult mongrel dogs were preconditioned to run voluntarily on a treadmill. We measured left ventricular pressure, aortic pressure and aortic flow. After complete recovery from the surgery, we imposed exercise stress of various degrees. The exercise stress varied stepwise in random order from a complete standstill to a maximum speed of (7 mph) with a 20% slope. We measured end-systolic elastance (Ees) of the left ventricle using a single beat estimation technique. Details of the technique have been described elsewhere (14-16). Briefly, we first predicted left ventricular isovolumic peak pressure by fitting a sinusoidal curve to its isovolumic portions. We then drew a tangential line to the left upper corner of the ventricular pressure vs ejected volume relationship. We estimated the ejected volume by integration of instantaneous aortic flow. With this technique, one could estimate Ees every beat without exact knowledge of the volume axis intercept of the end-systolic pressure-volume relationship. We obtained effective arterial elastance (Ea) by taking the ratio of end-systolic pressure to stroke volume (17).

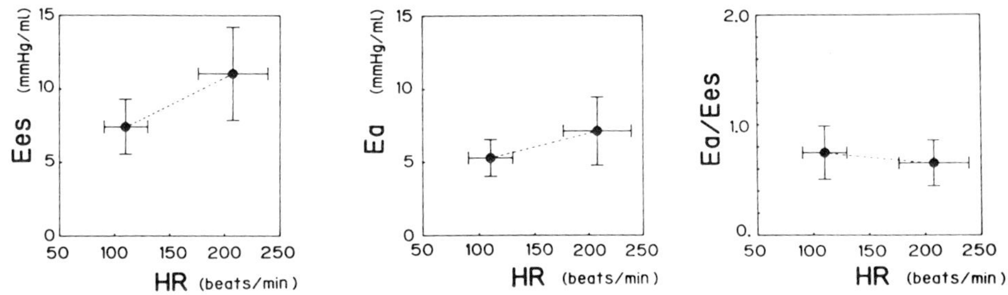

Fig. 1. The effect of exercise on Ees, Ea,and Ea/Ees. Both Ees and Ea increased with exercise. Ea/Ees, however, remained unchanged. Reprinted with permission from ref. 18.

Effect of Exercise on E_{es} and E_a

Exercise increased heart rate (108±17 vs 199±31 beats/min, $p<0.005$). At the same time it nearly doubled cardiac output (1.8±0.5 vs 3.51±1.1 l/min, $p<0.005$). Mean arterial pressure, however, increased marginally (77±4 vs 110±22 mmHg, $p<0.05$). Thus arterial resistance decreased by approximately 30% (2.8±0.6 vs 2.1±0.6 mmHg sec/ml, $p<0.05$). Figure 1 illustrates the effect of exercise on Ees and Ea. As shown in the left panel of Figure 1, Ees increased with heart rate (7.61±0.7 vs 10.9±2.6 mmHg/ml, $p<0.005$). The average slope of the Ees vs cardiac output relation was 2.10±0.64 mmHg/ml min/l. As shown in the middle panel of Figure 1, Ea also increased with heart rate (4.9±1.4 vs 6.7±1.8 mmHg/ml, $p=0.068$). The average slope of the cardiac output vs Ea relation was 1.40±0.66 mmHg/ml min/l. Since Ea approximates arterial resistance multiplied by heart rate, increases in heart rate overrode those in arterial resistance in determining Ea. The right panel of Figure 1 shows the effect of exercise on the ratio of Ea to Ees (Ea/Ees). As can be seen the ratio remains fairly constant (0.69±0.26 vs 0.63±0.21,ns) irrespective of changes in heart rate.

Optimality of Afterload: Maximum External Work

A given level of cardiac output can be achieved by various combinations of ventricular contractility, preload and afterload. Thus the regulatory system must choose a particular combination of these parameters. It is not clear, however, how optimal functioning of the cardiovascular system is naturally defined, whether in terms of cardiac output, stroke volume, external work or alternatively, efficiency of ventricular ejection. In view of the vital function of the cardiovascular system, performance would be more essential than efficiency, although the optimal loading conditions for performance and for efficiency may well be the same. We defined the optimal afterload as that at which external work of the left ventricle becomes maximum (18-20). With the framework of ventricular-arterial coupling (17), external work (EW) for quasi-isobaric ejecting contraction can be approximated as

$$EW = \frac{E_a}{(1+E_a/E_{es})^2}(V_{ed}-V_0)^2 \tag{1}$$

where V_{ed} is end-diastolic volume and V_o, systolic unstressed volume. EW becomes maximum (EW_{max}) when E_a equals E_{es}. Substituting E_a with E_{es} yields

$$EW_{max} = \frac{E_{es}}{4}(V_{ed}-V_0)^2 \tag{2}$$

We defined the optimality of the afterload (Q_{load}) as the ratio of external work to its maximum value (18-20), i.e.,

$$Q_{load} = \frac{EW}{EW_{max}} \tag{3}$$

Substituting EW and EW_{max} with Equations 1 and 2 yields

$$Q_{load} = \frac{4E_a/E_{es}}{(1+E_a/E_{es})^2} \tag{4}$$

Note that Q_{load} is the function of E_a/E_{es} and varies between 0 and unity. Q_{load} becomes unity (maximum) when E_a equals E_{es}. This is the condition at which external work becomes maximum. If Q_{load} is 0.5, the afterload extracts 50% of potential external work from a given left ventricle.

Optimality of Heart: Minimum Oxygen Consumption

We define the optimal left ventricle as the heart that requires minimum energy per unit time to support a peripheral demand (19,20). The peripheral demand is characterized by fixed cardiac output and fixed arterial resistance. Suga and his associates (21-23) demonstrated a tight linear correlation between the end-systolic pressure-volume area (PVA) and oxygen consumption (VO_2) (equivalent to metabolic demand) of the left ventricle. Changes in contractility affected the slope of the PVA-VO_2 relationship little, however, they altered the intercept significantly. The oxygen consumption per beat is expressed as

$$VO_2 = A \bullet PVA + B \bullet E_{es} + C \tag{5}$$

where A, B and C are experimentally determined constants. Equation 5 relates mechanical work of the left ventricle to its oxygen consumption. By combining the framework of ventricular-arterial coupling and that of the PVA-VO_2 relationship, we can estimate oxygen consumption of the heart for various combinations of contractility, preload and afterload. If end-diastolic volume is fixed, a unique combination of heart rate and Ees, which minimizes VO_2 to meet a peripheral demand, can be identified. We defined the optimality of the heart (Q_{heart}) as the ratio of minimum oxygen consumption (VO_{2min}) to the actual VO_2 per unit time (18,20), i.e.,

$$Q_{heart} = \frac{VO_{2min}}{VO_2} \tag{6}$$

Q_{heart} varies between zero and unity. The heart with Q_{heart} equal to 1 requires minimum oxygen to meet peripheral demand. If Q_{heart} is 0.5, the heart requires twice as much oxygen relative to the most efficient heart.

Effect of Exercise on Optimality of Ventricular Ejection

Figure 2 shows the effect of exercise on the optimality of ventricular coupling with the arterial system. The top panels show data from individual animals as a function of heart rate. The different symbols represent different animals. The bottom panels show the pooled data. As can be seen in the left panels, Q_{load} remained reasonably high irrespective of heart rate. The pooled data indicated that Q_{load} was 0.93 ± 0.07 at rest and 0.92 ± 0.08 at peak exercise. Thus external work was nearly maximum for a given contractility and preload at rest, and remained nearly maximum at peak exercise. The right panels of Figure 2 show the effect of exercise on Q_{heart}. As can be seen, Q_{heart} was close to unity indicating that the oxygen consumption of the left ventricle was nearly minimum during exercise as well as at rest for a given cardiac work. Thus it is fair to say that exercise hardly affected optimal ventricular coupling with the arterial system despite tremendous changes in hemodynamic conditions.

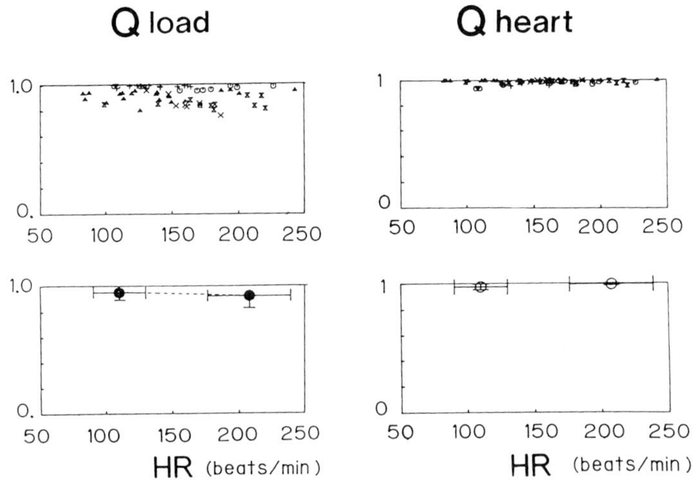

Fig. 2. The effect of exercise on the optimality of ventricular-arterial coupling. The top panels show individual data as a function of heart rate (HR). Different symbols represent different animals. The bottom panels show pooled data. Both optimality of load (Q_{load}) and heart (Q_{heart}) were close to unity. Exercise stress did not alter the optimal conditions (modified from Hayashida et al. (13)).

BAROREFLEX REGULATION OF EFFICIENCY OF VENTRICULAR-ARTERIAL COUPLING

What makes the coupling of the left ventricle with the arterial system optimum in terms of energy transmission is not well understood. Since the baroreflex plays a crucial role in regulating circulation, we investigated how the baroreflex system affects the optimality of the ventricular-arterial coupling (24). In particular, we analyzed the dynamic characteristics of the baroreflex regulation. This is because most of the stresses experienced in daily life would be predominantly dynamic rather than static.

Methods

In seven mongrel dogs, we anesthetized with sodium pentobarbital and exposed the carotid arteries. We ligated the internal and external carotid arteries bilaterally and isolated the carotid sinuses. We cannulated the carotid sinuses to allow the perturbation of pressure. We cut the vagus nerve bilaterally to remove the buffering effects of other baroreflex systems such as the aortic arch baroreflex and the cardiopulmonary baroreflex. We opened the chest, inserted a pressure transducer into the left ventricle and mounted an electromagnetic flow probe at the aortic root. We used the white-noise method to identify open-loop transfer functions from carotid sinus pressure to Ees and to Ea, because the white-noise method enables us to estimate the unbiased liner transfer function (25). Mean level of the carotid sinus pressure was matched to that of aortic pressure. The amplitude of the perturbation was 25 mmHg. We obtained Ees beat-by-beat with the single beat estimation technique. We obtained Ea by taking the ratio of end-systolic pressure to stroke volume.

Estimation of the Transfer Function

The transfer function is an expression of the linear input-output relation in the frequency domain (25-27). It represents the amplitude and phase relation between the input and output at each corresponding frequency. To obtain the transfer function from the carotid sinus pressure to Ees, we frequency-analyzed the time series of carotid sinus pressure, Ees and Ea. We then estimated the cross power spectrum between Ees and carotid sinus pressure. Dividing the cross power spectrum with the power spectrum of carotid sinus pressure yields the transfer function (H_{Ees}). Similarly, we estimated the cross power spectrum between the carotid sinus pressure and Ea. Dividing the cross power spectrum with the power spectrum of carotid sinus pressure yields the transfer function (H_{Ea}). We analyzed the transfer functions over the frequency range between 0.002-0.25 Hz.

Open Loop Transfer Functions From Carotid Sinus Pressure to E_{es} and to E_a

Averaged transfer functions are illustrated in Figure 3. The bold lines are means, and the two accompanying thin lines indicate SD. Note that both the frequency and modulus axes are logarithmically scaled. The left panel of Figure 3 illustrates H_{Ees}. The modulus was rather flat up to about 0.03 Hz and thereafter decayed at the rate of -40 db per decade. The phase was totally out of phase at lower frequencies and shifted up to slightly more than 180 degree at higher frequencies. These characteristics are consistent with a second-order delay system with a delay element. We calculated parameters for H_{Ees} fitting to the second-order delay system. The gain was 0.085 ± 0.065 with a natural frequency of 0.039 ± 0.013 Hz. This implied that the carotid sinus baroreflex mainly responded to changes in carotid sinus pressure in the frequency range below 0.039 Hz. The damping ratio was 0.65 ± 0.11. The right panel of Figure 3 is H_{ea}. It resembled H_{Ees} with a slight difference in phase characteristics. The gain was 0.081 ± 0.049, which was not significantly different from that of H_{Ees}. Neither the natural frequency (0.039 ± 0.007 Hz) nor damping ratio (0.64 ± 0.24) was different from that of H_{Ees}. Thus, the carotid sinus baroreflex dynamically altered Ees and Ea to the same extent in the process of stabilizing arterial pressure. As the net effect, the baroreflex will not affect the ratio of Ea to Ees.

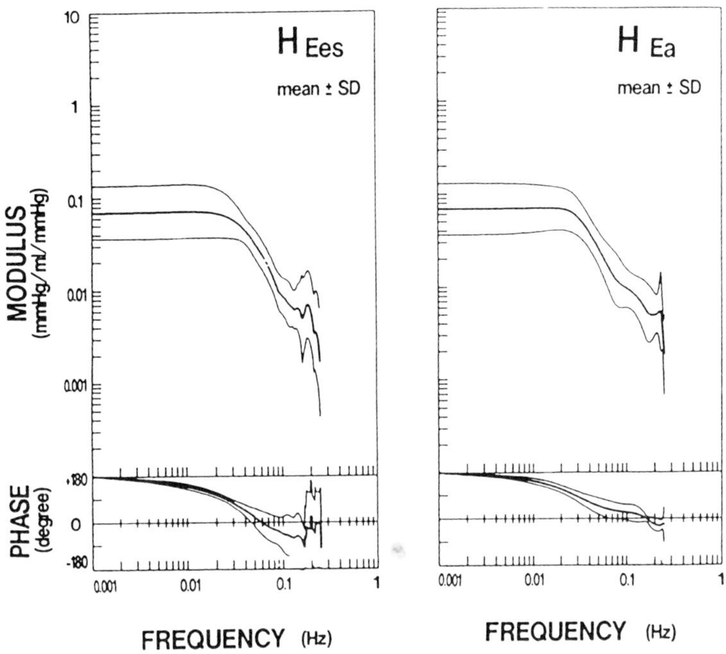

Fig. 3. Transfer functions from carotid sinus pressure to Ees (H_{Ees}) and to Ea (H_{Ea}). The similarity of the functions implies that dynamic adjustment of blood pressure through the baroreflex will not alter the ratio of Ea to Ees. Reprinted by permission from reference (24).

Effect of Baroreflex on Optimality of Ventricular-Arterial Coupling

Using these estimated transfer functions, we evaluated how the carotid sinus baroreflex affects ventricular-arterial coupling. Figure 4 illustrates a representative response of the simulation. In the control condition, Q_{load} is virtually unity. Q_{heart} is also close to unity. When intrasinus pressure abruptly decreased by 25 mmHg, the baroreflex increased arterial pressure. At the same time, both Ees and Ea increased. As anticipated, the time course of Ees and Ea were very close during the dynamic restoration process. Since the optimality of ventricular-arterial coupling is a function of the ratio of Ees to Ea, both Q_{load} and Q_{heart} remained close to unity over the entire time course of adjustment. The baroreflex system maintains arterial pressure without compromising the energy efficiency of ventricular-arterial coupling. The baroreflex system was preprogrammed to regulate the ventricular and arterial properties so as to optimize the coupling between them.

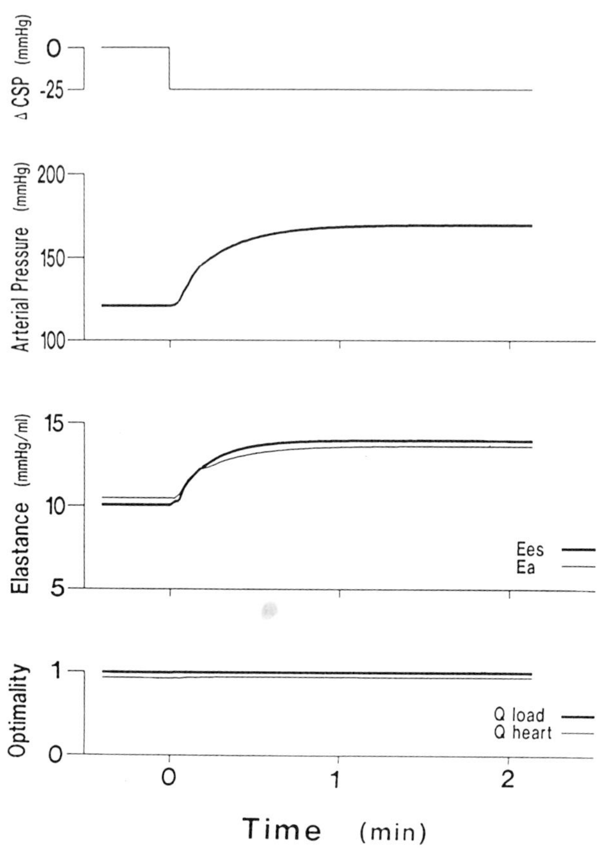

Fig. 4. Simulation of baroreflex regulation of arterial blood pressure. A sudden decrease in carotid sinus pressure (CSP) increases arterial pressure. Although it also increases both Ee and Ea, their values are reasonably close even in the transient state. Because of this balanced regulation of Ees and Ea, both Q_{load} and Q_{heart} hardly changed.

DISCUSSION

Stresses vs Optimality of Ventricular-Arterial Coupling and Baroreflex

We have shown that left ventricular external work was nearly maximum during exercise. The ventricle generates cardiac output to meet peripheral demand with minimum metabolic energy irrespective of the level of physical activity. Simultaneous optimization of the load and heart would be desirable for animals. This appears to be the case in the normal ventricular-arterial coupling. The transfer functions from carotid sinus pressure to Ees and to Ea were remarkably similar. In other words, the baroreflex changes both Ees and Ea similarly. As a result the baroreflex is capable of maximizing external work while minimizing the oxygen consumption of the ventricle. Although the mechanism by which the optimality of ventricular-arterial coupling is maintained is not well understood, the baroreflex regulation of ventricular-arterial coupling was in accordance with this physiological response.

The Ratio of E_a to E_{es} in Determining Optimalities of the Load and Heart

Figure 5 illustrates Q_{load} as a function of Ea/Ees. As anticipated Q_{load} becomes unity when Ea/Ees is unity. How, then, does the oxygen consumption of the heart relate to Ea and Ees? Using the framework of ventricular-arterial coupling and that of the PVA-VO_2 relation, one can derive the optimality of the heart as a function of Ea/Ees. As can be seen in Figure 5, Q_{heart} becomes unity when Ea/Ees is close to 0.5. If the ratio is below 0.5, Q_{load} decreases sharply. The fact that both Q_{load} and Q_{heart} become close to unity when Ea/Ees varies between 0.5 and 1.0, makes it possible to simultaneously maximize external work for a given preload and minimize oxygen consumption. Even when the dogs exercised, Ea/Ees remained within the narrow range (between 0.42 and 0.84). Again, in this range, both Q_{load} and Q_{heart} are close to unity. Whether these observations are intended results by the regulatory system or a simple coincidence remains unknown. Nevertheless, it is fair to say that the ventricular-arterial coupling of conscious dogs was very efficient from a view point of energy usage.

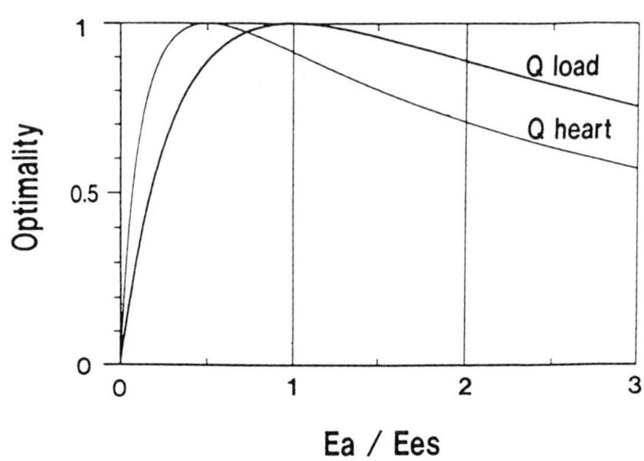

Fig. 5. Optimality of load (Q_{load}) and heart (Q_{heart}) as a function of Ea/Ees. When Ea/Ees varies between 0.5 and 1, external work of the ventricle is maximum while the oxygen consumption is minimum. Reprinted with permission from ref. 24.

Pathophysiological Significance

If ventricular contractility fails to increase in response to, for instance, exercise as in patients with heart failure, increased heart rate and increased resistance inevitably increases Ea and thus Ea/Ees. As shown in Figure 5, the increase in Ea/Ees beyond the normal range decreases both optimality of the load and heart. The deterioration of optimality of the load in the ventricle with poor contractility has been demonstrated in animals (12) and in patients (28). When we treat patients with heart failure, we often use positive inotropic agents and vasodilators. Positive inotropic agents with minimum chronotropic effects would decrease Ea/Ees, and, in turn, increase the optimality. In the case of pure vasodilators, the decrease in Ea will improve the optimality. When the left ventricle is hypertrophied due to increases in afterload, such as in patients with hypertension, Ees would increase (29). Since the increase in Ees is associated with an increase in arterial resistance, the resultant Ea/Ees may remain unchanged. This is supported by the fact that ejection fraction of patients with hypertension and hypertrophy are not necessarily low. Under this condition, optimality of the load is well maintained. Numeric analyses indicated that even if the increase in Ees is a consequence of the increased myocardial mass, Q_{heart} remains high as long as the inotropic state of the myocardium is unchanged. In case of hypertrophic cardiomyopathy without outflow obstruction, supernormal systolic function has been demonstrated (30). Since the increase in Ees is unrelated to the afterload, Ea/Ees would decrease. If ejection fraction is as high as 90%, the ventricle delivers only one-third of the potential external work. To support the same peripheral demand, such a heart requires 30% more oxygen than the normal heart. β-blockers or calcium channel antagonists depress systolic function, and thus, improves the optimality of the load and the ventricle. Hence, it seems that these treatments are reasonable from a viewpoint of the efficiency of energy usage.

CONCLUSION

We conclude that in normal dogs external work of the left ventricle is nearly maximum at rest and under moderate exercise. The metabolic energy required to generate cardiac output and to maintain arterial pressure to meet peripheral demand is nearly minimum as long as the heart is capable of responding to the command of the regulatory system. The baroreflex system appears to play a crucial role in maintaining arterial pressure without compromising energy efficiency of ventricular coupling with the arterial system.

REFERENCES

1. Piene, H., Sund, H. Flow and power output of right ventricle facing load with variable input impedance. Am J Physiol 237 (Heart Circ Physiol 6): H125-H130, 1979.

2. Elzinga, G., Piene, H., Jong, J.P. Left and right ventricular pump function and consequences of having two pumps in one heart. A study on the isolated cat heart. Circ Res 46:564- 574, 1980.

3. Piene, H., Sund, T. Does normal pulmonary impedance constitute the optimum load for the right ventricle? Am J Physiol 242:H154-H160, 1982.

4. Van den Horn, G.J., Westerhof, N., Elzinga, G. Optimal power generation by the left ventricle. A study in the anesthetized open thorax cat. Circ Res 56: 252-261, 1985.

5. Van den Horn, G.J., Westerhof, N., Elzinga, G. Feline left ventricle does not always operate at optimum power output. Am J Physiol 250 (Heart Circ Physiol 19):H961-H967, 1986.

6. Toorop, G.P., Gerardus, J., Van den Horn, G.J., Elzinga, G., Westerhof, N. Matching between feline left ventricle and arterial load:optimal power or efficiency. Am J Physiol 254 (Heart Circ Physiol 23):H279-H285, 1988.

7. Burkhoff, D., Sagawa. K. Ventricular efficiency predicted by an analytical model. Am J Physiol 250 (Regulatory Integrative Comp Physiol 19): R1021-R1027, 1986.

8. Tanaka, N,. Yasumura, Y., Nozawa, T., Futaki, S., Uenishi, M., Hiramori, K., Suga, H. Optimal contractility and minimal oxygen consumption for constant external work of heart. Am J Physiol 254 (Regulatory Integrative Comp Physiol 23): R933- R943, 1988.

9. Wilcken, D.E.L., Charlier, A.A., Hoffman, J.I.E., Guz, A. Effects of alterations in aortic impedance on the performance of the ventricles. Circ Res 14:283-293, 1964.

10. Sunagawa, K., Maughan, W.L., Sagawa, K. Optimal arterial resistance for the maximal stroke work studied in isolated canine left ventricle. Circ Res 56:586-595, 1985.

11. Suga, H., Igarashi, Y., Yamada, O., Goto, Y. Mechanical Efficiency of the left ventricle as a function of preload, afterload, and contractility. Heart and Vessel 1:3-8, 1985.

12. Myhre, E.S.P., Johansen, A., Bjornstad, Piene, H. The effect of contractility and preload on matching between the canine left ventricle and afterload. Circulation 73:161-171, 1986.

13. Hayashida, K., Sunagawa, K., Noma, M., Sugimachi, M., Ando, H., Nose, Y., Nakamura, M. Mechanical matching of the left ventricle with the arterial system in exercising dogs. Circ Res 71:481-489, 1992.

14. Sunagawa, K., Yamada, A., Senda, Y., Kikuchi, Y., Nakamura, M., Shibahara, T., Nose, Y. Estimation of the hydromotive source pressure from ejecting beats of the left ventricle. IEEE BME-27:299-305, 1980.

15. Sunagawa, K., Sagawa, K. Models of ventricular contraction based on time-varying elastance. CRC Critical Rev Biomed Eng 7:193-228, 1982.

16. Takeuchi, M., Igarashi, Y., Tomimoto, S., Odake, M., Hayashi, T., Tsukamoto, T., Hata, K., Takaoka, H., Fukzaki, H. Single beat estimation of the slope of the end-systolic pressure-volume relation in the human left ventricle. Circulation 83:202-212, 1991.

17. Sunagawa, K., Maughan, W.L., Burkhoff, D., Sagawa, K.. Left ventricular interaction with arterial load studied in isolated canine ventricle. Am J Physiol 245 (Heart Circ Physiol 14):H773-H780, 1983.

18. Sunagawa, K., Hayashida, K., Sugimachi, M., Noma, M., Ando, H., Tajimi, T., Tomoike, H., Nose, Y., Nakamura, M. Ventriculoarterial matching in exercising dogs. In: Sideman, S., Beyer, R. (ed) Analysis and simulation of the cardiac system-ischemia. CRC Press Boca Raton, pp 89-98, 1989.

19. Sunagawa, K., Sugimachi, M., Todaka, K., Nakamura, M. Ventricular matching with the arterial system in chronically instrumented dogs. In: Hori, M., Suga, H., Baan, J., Yellin, E.L. (ed) Cardiac mechanics and function in the normal and diseased heart. Springer Tokyo, pp 207-210, 1989.

20. Sugimachi, M., Todaka, K., Sunagawa, K., Nakamura, M. Optimal afterload for the heart vs optimal heart for the afterload. Frontiers Med Biol Eng 2:217-221, 1990.

21. Suga, H. Total mechanical energy of a ventricle model and cardiac oxygen consumption. Am J Physiol 236 (Heart Circ Physiol 5):H498-H505, 1979.

22. Suga, H., Hayashi, T., Shirahata, M. Ventricular systolic pressure-volume area as predictor of cardiac oxygen consumption. Am J Physiol 240 (Heart Circ Physiol 9):H39-H44, 1981.

23. Suga, H., Yamamura, Y., Nozawa, T. Prospective prediction of O2 consumption from pressure volume area (PVA) in dog heart. Am J Physiol 252 (Heart Circ Physiol 21):H1258-H1268, 1987.

24. Kubota, T., Alexander J., Jr, Itaya, R., Todaka, K., Sugimachi, M., Sunagawa, K., Nose, Y., Takeshita, A. Dynamic effects of carotid sinus baroreflex on ventriculo-arterial studied in anesthetized dogs. Circ Res 70:1044-1053, 1992.

25. Marmarelis, P.Z., Marmarelis, V.Z. Analysis of physiological system: The white noise approach. Plenum Press, New York, 1978.

26. Marple, S.L. Digital spectral analysis with applications. Prentice-Hall, New Jersey, 1987.

27. Johansson, R. System modeling and identification. Prentice-Hall, New Jersey, 1993.

28. Asanoi, H., Sasayama, S., Kameyama, T. Ventriculoarterial in normal and failing heart in humans. Circ Res 65:483-493, 1989.

29. Sasayama, S., Franklin, D., Ross, J., Jr. Hyperfunction with normal inotropic state of the hypertrophied left ventricle. Am J Physiol 232 (Heart Circ Physiol):H418-H422, 1977.

30. Raizner, A.E., Chahine. R.A., Ishimon, T., Audec, M. Clinical correlates of left ventricular cavity obliteration. Am J Cardiol 40:303-309, 1977.

13. Ventriculo-Arterial Load Matching of Failing Hearts

Hidetsugu Asanoi, Tomoki Kameyama, and Shinju Ishizaka

INTRODUCTION

Katz proposed that heart failure should be regarded as a cardiomyopathy of overload which leads to an energy-starved state and progressive deterioration (1). Therefore, effective transfer of mechanical energy of ventricular contraction crucially affects the long-term outcome of chronic heart failure. To apply this concept to heart failure management in clinical settings, energetically optimal ventriculo-arterial coupling should be quantitatively defined. To date, a number of experimental and clinical studies(2-6) have been performed to answer the question: what is the energetically optimal coupling between the ventricle and arterial load actually operative in vivo under physiological and pathological circumstances?

VENTRICULO-ARTERIAL COUPLING FRAMEWORK

Criteria for Optimal Coupling

From an engineering point of view, the ventricle is considered to be a generator of hydraulic energy, which transfers the mechanical energy of contraction to the blood accumulated in the ventricular chamber. The stroke work (SW) and power output from a given end-diastolic volume are strongly influenced by the input impedance of the arterial system. In physics and engineering, an energy source and its load are considered matched when a maximal amount of energy is transferred from the source to the load. Matching occurs when the input impedance of the load equals the output impedance of the energy source. Several investigators(2-6) have confirmed the validity of the concept that the physiological state of the ventricle and arterial input impedance are matched to produce a maximal SW or power. Another criterion for optimal coupling between an energy source and its load is the principle of economical fuel consumption or mechanical efficiency, defined as the ratio of SW to myocardial oxygen consumption per beat.

Theoretically Maximal Stroke Work and Mechanical Efficiency

The analytical framework developed by Sunagawa et al(3,7) and the pressure-volume area(PVA) concept proposed by Suga et al(8,9) allow prediction of the conditions for maximal left ventricular SW or mechanical efficiency. Within this framework ventricular contractility is quantified by the slope(Ees) of the end-systolic pressure-volume(P-V) relationship, and the arterial input impedance is expressed by the slope(Ea) of the end-systolic pressure-stroke volume relationship. With increases in Ea, SW initially increases, reaches a plateau, and then decreases. As shown by Sunagawa et al(3), maximum SW occurs when arterial and ventricular properties are equalized(Ea=Ees). Ventricular efficiency, defined as the ratio of external SW to myocardial oxygen consumption per beat, initially rises with increases in Ea, reaches a maximum, and then decreases. Burkhoff and Sagawa(10) theoretically predicted that maximal efficiency occurs when Ea is less than Ees (Ea=Ees/2). This conceptual framework indicates that energetically optimal ventriculo-arterial coupling should exist when the Ea/Ees ratio ranges from 0.5 to 1.0 and that the afterload that results in the greatest efficiency is always less than that which provides the maximum SW.

VENTRICULO-ARTERIAL COUPLING IN HUMANS

The concept of ventriculo-arterial interaction has been extended to human studies, and physiologic matching of cardiac performance with arterial load has been evaluated in terms of the above-mentioned ventricular and arterial volume elastance(6,11). Eight normal subjects with no signs and symptoms of cardiac disease, 4 patients with atypical chest pain, and 16 patients with cardiac dysfunction constituted the study group. All patients had supporting clinical, chest radiographic, and echocardiographic evidence of impaired left ventricular function. Severity of cardiac disease ranged from Class II to Class III. Subjects were divided into three groups based on their resting left ventricular ejection fraction (EF). Group A consisted of 12 subjects with left ventricular EF of $\geq$60%. Group B consisted of seven patients with mild left ventricular dysfunction in whom EF was 40-59%. Group C consisted of nine patients with greater left ventricular dysfunction, in whom EF was <40%.

Figure 1 illustrates representative ventricular end-systolic P-V relationships and arterial end-systolic pressure-stroke volume relationships for patients in groups A, B, and C. Linear relationships between corresponding end-systolic pressure and volume have been observed in these patients. Ees was 4.5 mmHg/ml/m^2 in group A and this value decreased substantially as the heart failed (2.5 mmHg/ml/m^2 in group B and 1.5 mmHg/ml/m^2 in Group C). With a reduction of EF, left ventricular end-diastolic volume was progressively augmented and Ea tended to increase. In group A, Ea/Ees was 0.50, with Ees being nearly twice as much as Ea. In group B, Ea/Ees was 0.97, with Ees being nearly equal to Ea. In group C, Ea/Ees increased to 2.29, with Ees being far less than Ea. Left ventricular pump efficiency, indicated by the ratio of SW to PVA, was highest in Group A (82±5%) and progressively fell with a reduction of EF (to 69±5% in Group B and 50±10% in Group C). SW was near its maximum in Group B patients with moderate cardiac failure, with the ventricle providing maximal transfer of mechanical energy of contraction to the arterial system.

Thus, the normal coupling condition in Group A is surprisingly similar to the data of Burkhoff and Sagawa(10), and cardiovascular interaction in normal subjects was set to optimize left ventricular mechanical efficiency in the resting state. Moderately depressed hearts maximize SW as demonstrated by Sunagawa(3) but they

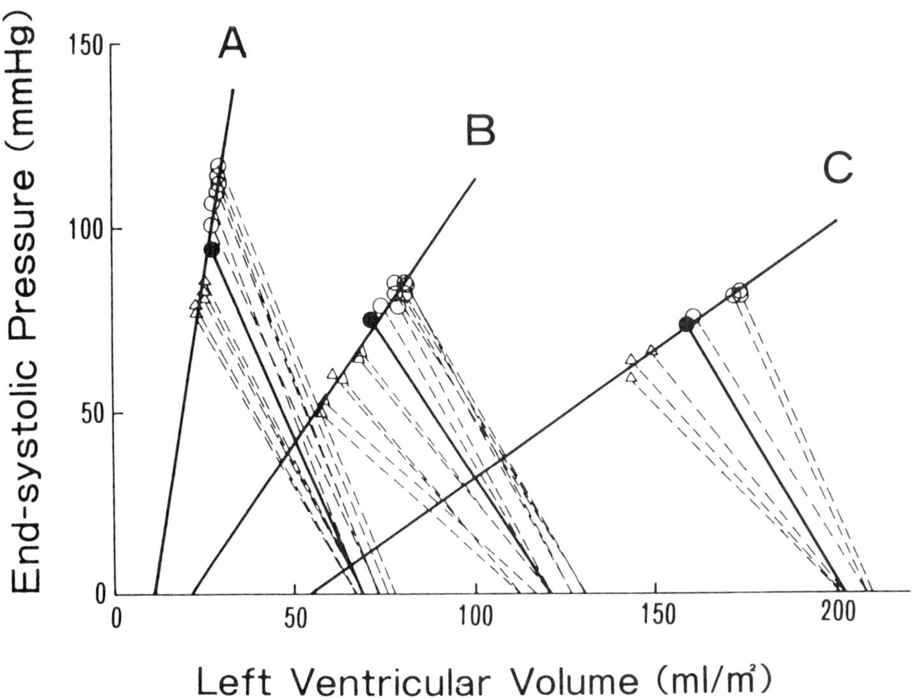

Fig. 1. Representative ventricular end-systolic pressure-volume and arterial end-systolic pressure-stroke volume relationships for patients in groups A, B, and C, at baseline(solid lines) and after pressure manipulations(dashed lines). Reprinted with permission from ref. 11.

generate a greater potential energy with consequent reduction in pump efficiency. Severely depressed hearts were no longer capable of maintaining SW or mechanical efficiency properly. Under all circumstances, however, arterial blood pressure remained essentially the same and severe hypotension was uncommon in chronic heart failure. These findings indicate that the deterioration of energy transfer in heart failure might therefore result from cardiovascular adjustment to maintain arterial pressure within the physiological range.

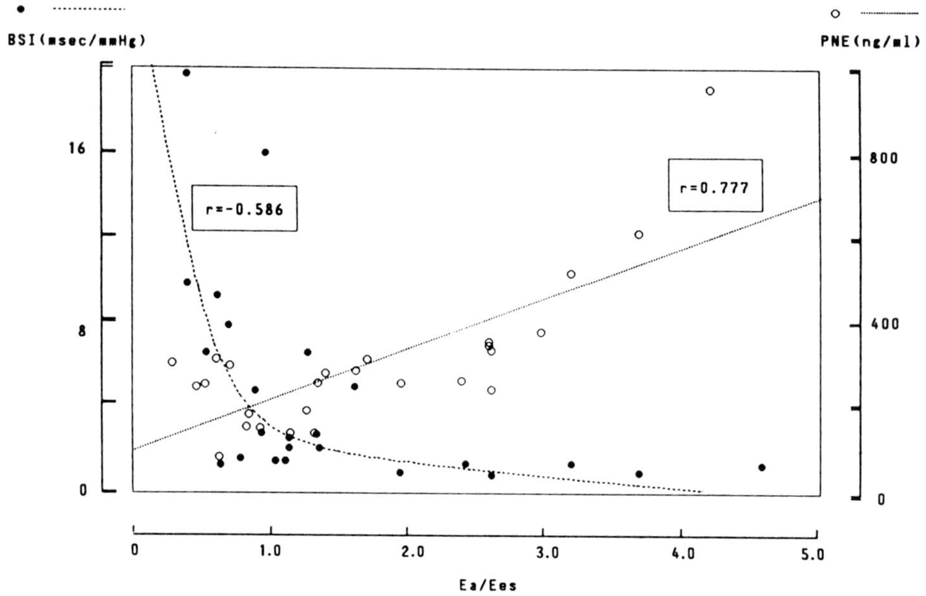

Fig. 2. Relationship between neurohumoral factors and ventriculo-arterial coupling. BSI=arterial baroreflex sensitivity index, PNE=plasma norepinephrine concentration.

NEURAL MODULATION OF VENTRICULO-ARTERIAL COUPLING

In heart failure neurohumoral mechanisms play the most important role in maintaining normal arterial pressure. The augmentation of sympathetic nervous activity serves to ensure cardiovascular homeostasis initially, but in later phases it may actually impede left ventricular performance through excessive peripheral vasoconstriction. Thus, as heart failure develops, matching between the ventricle and the arterial system would be modulated by changes in neurohumoral mechanisms such as sympathetic and parasympathetic nervous activities. Although the sympathetic nervous system has been confirmed to be overly active in congestive heart failure, there is limited information concerning quantitative analysis of the relationships between sympathetic activity and ventricular-load interaction.

Clinical Study

To relate the ventriculo-arterial coupling to plasma norepinephrine and baro-reflex sensitivity and to grade the stage of circulatory adjustment to heart failure, 10 normal subjects and 23 patients with cardiac dysfunction were studied. Figure 2 illustrates the relation between ventriculo-arterial coupling as defined by the ratio of Ea/Ees and the neurohumoral factors. Both baroreflex sensitivity and plasma norepinephrine concentration correlated well with Ea/Ees. The upper limit of the normal Ea/Ees ratio was 1.0, beyond which baroreflex sensitivity substantially fell, whereas plasma norepinephrine level was markedly increased in patients with Ea/Ees ratio of 2.0 or more(12). Our data suggest that in the course of heart failure circulatory adjustments result in three phases. When Ea/Ees is 0.5 or less, the arterial baroreflex is preserved with little sympathetic activation. In contrast, when Ea/Ees is 1.5 or more the baroreflex is markedly depressed, with enhanced sympathetic activity. Within the Ea/Ees range of 0.5 to 1.5, where SW can be maximized, both parasympathetic and sympathetic neural function seems to be balanced.

Experimental Study

Despite accumulating data on ventriculo-arterial coupling in various situations, the feedback mechanisms responsible for SW optimization remain to be delineated. To investigate the role of autonomic reflexes, ventriculo-arterial coupling was examined under autonomically intact and blocked conditions in 8 trained, conscious dogs instrumented with micromanometer and conductance volume catheter in the left ventricle(13). Left ventricular pressure and volume were recorded simultaneously throughout the cardiac cycle. Pneumatic cuffs around the superior and inferior venae cavae were inflated transiently to decrease peak left ventricular systolic pressure by about 30 mmHg. The left ventricular end-systolic P-V relation was then analyzed from the superimposed series of P-V loops during this pressure change. A previous study documented that the end-systolic P-V curve could be obtained reproducibly in this model(14). After baseline measurements, an intravenous infusion of sodium nitroprusside was started to decrease left ventricular end-systolic pressure by approximately 25 mmHg. When a stable reduction in peak left ventricular pressure was attained, caval occlusion was repeated. The nitroprusside infusion was then stopped. After pressures returned to control levels, angiotensin II was infused intravenously in five dogs to increase end-systolic pressure by about 30 mmHg. The non-neuronal contribution to ventriculo-arterial coupling was examined while autonomic reflexes were blocked by intravenous bolus injections of 0.2 mg/kg of atropine and 6.0 mg/kg of hexamethonium chloride followed by continuous infusion of both agents (1 mg/hr atropine and 3 mg/kg/hr hexa-methonium). When left ventricular pressure and heart rate were stabilized, left ventricular pressure and volume were again recorded serially during caval occlusions before and after the administration of vasoactive agents.

Ventriculo-Arterial Coupling with Reflexes Intact or Blocked

Figure 3 illustrates left ventricular end-systolic P-V relations before and during vasoactive agent treatment with the autonomic system intact. Nitroprusside substantially increases Ees with little change in Ea, resulting in a fall of the Ea/Ees ratio. In contrast, angiotensin II increases the Ea/Ees ratio from baseline due to an increase in Ea while Ees did not change appreciably. Left ventricular pressure-volume relations under autonomic blockade are shown in Figure 4. Ees decreases with nitroprusside and increases with angiotensin II in parallel with changes in Ea, so that the Ea/Ees ratio does not vary from the baseline value as much as it does with the autonomic system intact. These results indicate that under autonomically blocked conditions the Ea/Ees ratio did not change from its baseline during treatment with vasoactive agents as much as it did with reflexes intact.

Variability of Ventriculo-Arterial Coupling

Generally, the body monitors the adequacy of the circulation by sensing arterial pressure and responds in a stereotypical manner to maintain arterial pressure within a preset range (14). This control is mediated largely through the autonomic nervous system, which innervates not only the heart but also the vascular system. Recently, Kubota et al (15) have evaluated dynamic effects of the carotid sinus reflex on ventriculo-arterial coupling and demonstrated that the baroreflex changes Ees and Ea almost to the same extent to optimize stroke work when the left ventricle and arterial system are intact. However, when responsiveness of either ventricle or arterial

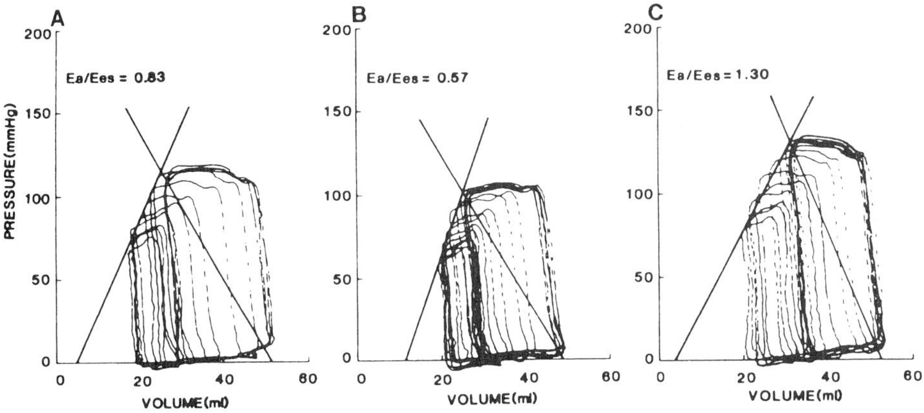

Fig. 3. Left ventricular end-systolic pressure-volume relationship before(A) and during vasoactive agent treatment with autonomic system intact. Nitroprusside(B) increases Ees with little change in Ea, resulting in a fall in Ea/Ees. In contrast, angiotensin II (C) increases Ea/Ees from baseline due to an increase in Ea, whereas Ees did not change. Reprinted with permission from ref. 13.

system is suppressed as during the use of vasoactive agents, or in cases of depressed cardiac contractility or limited preload reserve, the unaffected part of the cardiovascular system must bear the brunt of responding to the body's requirements.

Under such circumstances, the cooperative adaptation of ventricle and arterial system can be lost. Thus, negative feedback from the autonomic nervous system to maintain normal arterial pressure and the ability of the cardiovascular system to respond to the baroreflex would determine variability in ventriculo-arterial coupling in animals with intact autonomic systems.

ENERGETICALLY OPTIMAL ARTERIAL PRESSURE

Effects of Loading and Inotropic Conditions on Mechanical Efficiency

In patients with severely depressed hearts, the working point of the left ventricle was shown to deviate from both optimal SW and mechanical efficiency because of cardio-

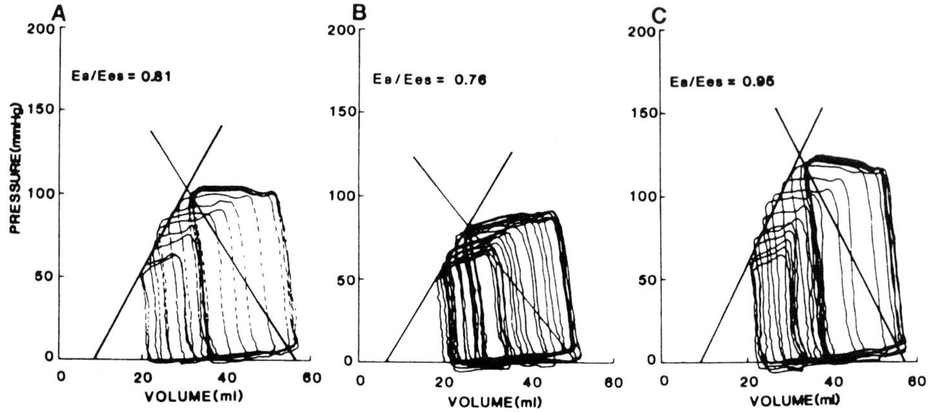

Fig. 4. Left ventricular end-systolic pressure-volume relationship before(A) and during vasoactive agent treatment under autonomic blockade. Ees decreases with nitroprusside (B) and increases with angiotensin II (C) in parallel with changes in Ea, so that Ea/Ees does not vary from baseline as much as it does with autonomic system intact. Reprinted with permission from ref. 13.

vascular adjustment to maintain arterial pressure within the physiological range. Therefore, an energetically optimal arterial pressure should exist for the failing heart. On the basis of the PVA concept, Suga et al(9) demonstrated in excised canine left ventricle that mechanical efficiency of the left ventricle varies as an explicit function of preload, afterload, and contractility. That is, the theoretically attainable maximal efficiency increases with increases in preload (end-diastolic volume). Similar relations are observed between mechanical efficiency and ventricular afterload pressure and between mechanical efficiency and contractility. This function allows us to predict mechanical efficiency of the failing left ventricle as well as to search for the optimal arterial pressure to maximize the mechanical efficiency of the left ventricle.

Optimal Arterial Pressure for Failing Hearts

The end-systolic pressure to maximize mechanical efficiency was predicted for a given end-diastolic volume and contractility to quantitatively define how far actual arterial pressure deviates from energetically optimal arterial pressure in 13 patients with cardiac dysfunction. Left ventricular SW, PVA, and Ees were obtained by contrast ventriculography at baseline and during pressure loading. Myocardial oxygen consumption was measured at the same time using thermodilution flow and oxygen difference from a coronary sinus catheter. Left ventricular end-systolic P-V relation and oxygen consumption-PVA relation were determined in each patient(17). In patients with less compromised ventricular function(Ees$\geq$3mmHg/ml) operating end-systolic pressure was close to the optimal pressure(109$\pm$41mmHg), achieving nearly maximal mechanical efficiency. As the heart deteriorated(Ees$<$3mmHg), however, the optimal end-systolic pressure(39$\pm$22mmHg) became significantly lower than the normal level while actual pressure was maintained within the normal range. This discrepancy resulted in worsening of the ventriculo-arterial coupling and mechanical efficiency from the viewpoint of maximal efficiency.

Therapeutic Implications

These findings suggest two strategies to optimize mechanical efficiency of the failing left ventricle. One is to reduce operating arterial pressure toward the optimal pressure without changing contractility. This corresponds to the use of vasodilators. Under circumstances where pressure-raising mechanisms are strongly activated, drugs which directly interfere with these systems, angiotensin-converting enzyme inhibitors or beta-blockers, seem to be more reasonable to decrease arterial pressure toward the optimal value for ventriculo-arterial coupling. Another strategy to optimize mechanical efficiency of the failing heart is to increase the optimal arterial pressure toward the normal range. This corresponds to the use of inotropic agents.

Augmentation of contractility could optimize mechanical efficiency of the failing heart while maintaining normal arterial pressure and potentially attenuating endogenous neurohumoral activity.

VENTRICULAR-LOAD OPTIMIZATION WITH MEDICAL TREATMENT

Effects of Vasodilator and Inotropic Agent

The observations we have previously shown raise the question whether ventricular-load mismatch can be restored from the viewpoint of optimal coupling when the heart or periphery is influenced by medical treatment. In 22 patients with heart failure, the role of preload, afterload, or contractility in ventricular-load interaction was independently examined in terms of optimal coupling(18,19). Lower body negative pressure (LBNP) was applied to reduce preload and nitroprusside was used

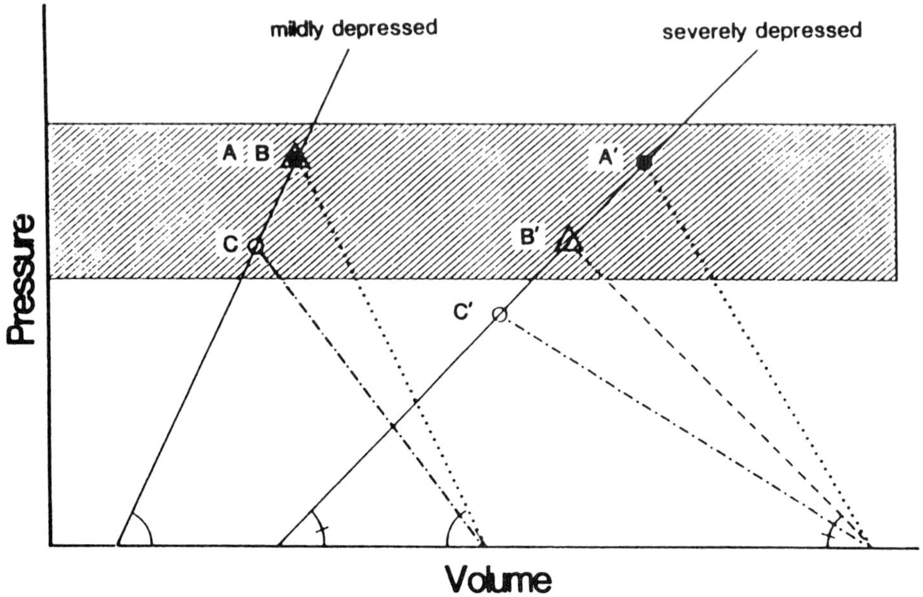

Fig. 5. Schematic representation of the effect of afterload reduction on ventricular-arterial coupling. The shaded area represents normal range of arterial pressure. In patients with mildly depressed hearts, the Ea/Ees ratio is nearly one in the baseline state (A), providing maximal stroke work (B). With afterload reduction, ventricular efficiency is augmented within the normal range of systemic pressure (C). In patients with severely depressed hearts, Ea is always greater than Ees in the baseline state (A'), with smaller stroke work than when Ea is equal to Ees (B'). Afterload reduction achieves maximal stroke work for a given contractility and preload (B') but cannot maintain the level of arterial pressure ideal for efficiency matching (C') because efficiency matching always occurs at a lower end-systolic pressure than that for stroke work matching. Reprinted with permission from ref. 18.

to decrease afterload. Dobutamine was used to enhance inotropic state. Before nitroprusside infusion, baseline Ea/Ees was 1.96 ± 1.08, such that the heart could maximize neither external work nor mechanical efficiency. Nitroprusside significantly decreased Ea/Ees to 1.45 ± 0.77 by largely reducing Ea, whereas LBNP increased the ratio to 2.37 ± 1.17. Dobutamine changed Ea/Ees from 1.43 ± 0.57 to 0.82 ± 0.47 by increasing Ees. Thus, left ventricular preload, afterload, and contractility could affect ventriculo-arterial coupling in a different manner. Reduction in afterload (rather than preload), or augmentation of contractility partially restored the disturbed ventricular-load coupling in heart failure toward the optimal conditions. The capability of vasodilators to optimize stroke work or mechanical efficiency seems to depend on the baseline inotropic state in each patient. If we assume that ventricular contractility remains unchanged with vasodilators, efficiency is always maximized at a lower level of end-systolic pressure than the pressure for external work to be maximized. Therefore, a severely depressed heart (Ea> >Ees) cannot maintain adequate systemic circulation at the level of arterial pressure for optimal efficiency(Figure 5,C'). In such cases, concomitant use of an inotropic agent may help maintain normal arterial pressure as well as optimal mechanical efficiency. On the other hand, in patients with less compromised cardiac function whose Ees is nearly equal to Ea, optimal efficiency could be attained by afterload reduction without deterioration of the systemic circulation(Figure 5,C). This concept might provide a rational basis for preventive use of vasodilators in patients with asymptomatic cardiac dysfunction.

CONCLUSIONS

As the heart fails, ventriculo-arterial coupling becomes deviated from matched conditions for maximal SW or mechanical efficiency. This ventriculo-arterial mismatching is closely related to neurohumoral abnormalities seen in patients with heart failure. A theoretical analysis using left ventricular and arterial elastance and the PVA concept demonstrated that mechanical efficiency in failing human heart is always maximized at a lower arterial pressure than the normal level, whereas operating arterial pressure remains maintained within the normal range in these patients. These observations suggest that feedback neural regulation to restore normal arterial pressure leads the failing heart to deviate from energetically optimal conditions and that agents which interfere with this mechanism would be promising in the treatment of chronic heart failure.

ACKNOWLEDGEMENTS

The authors are deeply indebted to Prof. Shigetake Sasayama for his continuous encouragement and valuable suggestions.

REFERENCES

1. Katz, A.M. Cardiomyopathy of overload: A major determinant of prognosis in congestive heart failure. N Engl J Med 322:100-110, 1990.

2. Van den Horn G.J., Westerhof N., Elzinga G. Optimal power generation by the left ventricle: a study in the anesthetized open thorax of a cat. Circ Res 56:252-261, 1985.

3. Sunagawa K., Maughan W.L., Sagawa K. Optimal arterial resistance for the maximal stroke work studied in isolated canine left ventricle. Circ Res 56:586-595, 1985.

4. Toorop G.P., Van den Horn G.J., Elzinga, G., Westerhof N. Matching between feline left ventricle and arterial load: optimal external power or efficiency. Am J Physiol 254:H287-H295, 1988.

5. Little, W.C., Cheng C.P. Left ventriculo-arterial coupling in conscious dogs. Am J Physiol 261:H70-H76, 1991.

6. Asanoi, H., Sasayama S., Kameyama T. Ventriculo-arterial coupling in normal and failing heart in humans. Circ Res 65:483-493, 1989.

7. Sunagawa K., Maughan W.L., Burkhoff D., Sagawa, K. Left ventricular interaction with arterial load studied in isolated canine ventricle. Am J Physiol 245:H773-H780, 1983.

8. Suga H., Hayashi, T., Shirahata M., Suehiro s., Hisano, R. Regression of cardiac oxygen consumption on ventricular pressure-volume area in dog. Am J Physiol 240:H320-H325, 1981.

9. Suga, H., Igarashi, Y., Yamada, O., Goto, Y. Mechanical efficiency of the left ventricle as a function of preload, afterload, and contractility. Heart & Vessels 1:3-8, 1985.

10. Burkhoff, D., Sagawa K. Ventricular efficiency predicted by an analytical model. Am J Physiol 250:R1021-R1027, 1986.

11. Sasayama, S., Asanoi, H. Coupling between the heart and arterial system in heart failure. Am J Med 90(5B):14-18, 1991.

12. Asanoi, H., Sasayama, S. Relationship of plasma norepinephrine to ventricular-load coupling in patients with heart failure. Jpn Circ J 53:131-140, 1989.

13. Asanoi, H., Ishizaka, S., Kameyama T., Sasayama, S.: Neural modulation of ventriculo-arterial coupling in conscious dogs. Am J Physiol 266:H741-H748, 1994.

14. Asanoi, H., Ishizaka S., Kameyama, T., Nozawa, T., Miyagi, M., and Sasayama, S. Serial reproducibility of conductance catheter volumetry of left ventricle in conscious dogs. Am J Physiol 31:H911-H915, 1992.

15. Harris, P. Congestive heart failure: central role of the arterial blood pressure. Br Heart J 58:190-203, 1987.

16. Kubota, T., Alexander, J., Itaya, R., Todaka, K., Sugimachi, M., Sunagawa, K., Takeshita, A. Dynamic effect of carotid sinus baroreflex on ventriculo-arterial coupling studied in anesthetized dogs. Circ Res 70:1044-1053, 1992.

17. Kameyama, T., Asanoi, H., Ishizaka, S., Yamanishi, K., Fujita, M., Sasayama, S. Energy conversion efficiency in human left ventricle. Circulation 85:988-996, 1992.

18. Kameyama, T., Asanoi, H., Ishizaka, S., Sasayama, S. Ventricular-load optimization by unloading therapy in patients with heart failure. J Am Coll Cardiol 17:199-207, 1991.

19. Ishizaka, S., Asanoi, H., Kameyama, T., Sasayama, S. Ventriculo-arterial optimization by inotropic stimulation in patients with heart failure. Int J Cardiol 31:51-58, 1991.

14. Calcium Signaling and Pharmacology of Cardiotonic Agents

Masao Endoh

INTRODUCTION

The mechanisms of action of drugs that act on myocardial contractility through different signal transduction pathways have lately been elucidated to a great extent, in particular since the introduction of novel methods for detecting changes in the intracellular concentration of Ca^{++} ions, ($[Ca^{++}]_i$), by the application of aequorin, quin-2, fura-2 and indo-1 in intact, contracting myocardial cells. It has been elucidated unequivocally that the regulation of myocardial contractility induced by cardiotonic agents is achieved either by increasing the mobilization of intracellular Ca^{++} ions via divergent subcellular mechanisms, by modulation of the responsiveness of myofibrils to Ca^{++} ions or by a combination of these two mechanisms. The mechanisms of action of drugs on myocardial contractility are classified into three processes (1). First, the binding of Ca^{++} ions to troponin C plays a central role by triggering a series of biochemical and mechanical processes in myofibrils which lead to sliding of myofilaments or force development, a process which is termed the central mechanism. Some drugs, such as caffeine (2) or agonists for β-adrenoceptors (3) act on the central mechanism, i.e., by respectively increasing or decreasing the affinity for troponin C to bind Ca^{++} ions. Second, the regulation of mobilization of intracellular Ca^{++} ions is a process upstream to the central mechanism and is termed the upstream mechanism. The upstream mechanism is the main process that is influenced by a number of known inotropic agents, including cardiac glycosides, catecholamines, Ca^{++} agonists and Ca^{++} antagonists. Third, modulation of the process subsequent to the binding of Ca^{++} ions to troponin C, namely, cross-bridge cycling is accomplished via the downstream mechanism. In intact myocardial cells, the upstream mechanism is easily distinguished from the latter two mechanisms by simultaneous measurement of the signals of intracellular Ca^{++} indicators and contractile force with application of the newly developed methods. But it is difficult to differentiate the downstream from the central mechanism. One has to confer with the findings of *in vitro* studies, including skinned cardiac fibers, determination of the activity of actomyosin ATPase and motility assays that make it possible to investigate the interaction of actin and myosin molecules in the absence of regulatory Ca^{++} ions (4).

While a number of drugs affect the cardiac inotropic response, only a few subcellular signal-transduction pathways have been found to be responsible for the drug-induced inotropic regulation. The purpose of this chapter is to focus on the characteristics of mechanisms of action of cardiotonic drugs, acting through three different subcellular mechanisms, namely, via accumulation of cyclic AMP, via acceleration of the hydrolysis of phosphoinositides, and via an increase in myofibrillar responsiveness to Ca^{++} ions, namely, the central and/or downstream mechanism(s).

CA^{++} TRANSIENTS

The rapid, transient increase in $[Ca^{++}]_i$ that occurs prior to the generation of force plays a central role in excitation-contraction (E-C) coupling in intact cardiac muscle (1,5). This transient increase in $[Ca^{++}]_i$, subsequent to excitation, is termed the Ca^{++} transient and it is an essential determinant of the time course of cardiac contraction and relaxation in twitch contraction. Alterations of the extracellular Ca^{++} ion concentration, $([Ca^{++}])_o$, modulate the amplitude of Ca^{++} transients and force of contraction essentially in parallel. A linear relation is obtained between the peak Ca^{++} transients and force. It is thought that the alteration of $[Ca^{++}]_o$ affects the cardiac E-C coupling via the upstream mechanism in a direct manner. Although the relation between the peak Ca^{++} transients and force does not always reflect the responsiveness of contractile proteins to Ca^{++} ions when the time course of Ca^{++} transients is changed (6,7), it provides a useful method for analysis of the mechanism of action of drugs if the factors involved (particularly the changes in duration of Ca^{++} transients) are carefully taken into consideration for analysis of the data (8).

The amplitude and duration of Ca^{++} transients are determined by several transmembrane and intracellular events that are pertinent to mobilization and handling of intracellular Ca^{++} ions, including the actions of voltage-dependent L-type Ca^{++} channels, Na^+-Ca^{++} exchange and the Ca^{++} pump of the sarcolemma, the Ca^{++} pump and Ca^{++}-release channels of the sarcoplasmic reticulum (SR), calcium-binding proteins, such as calmodulin, and mitochondria. In particular, the release of Ca^{++} ions from the SR makes the largest contribution to the amplitude of Ca^{++} transients in mammalian cardiac muscle (5). The uptake of Ca^{++} ions by the SR, mediated by the Ca^{++} pump, and the binding of Ca^{++} ions to and their release from troponin C play an important role in determining the shape of Ca^{++} transients (1).

A number of inotropic drugs and interventions modulate the upstream mechanisms. The overall regulatory mechanism of individual inotropic interventions in intact myocardial cells can be analyzed by simultaneous assessment of the strength and the time course of twitch contractions and Ca^{++} transients (7-10). Biochemical and electrophysiological findings, as well as those from experiments with skinned cardiac fibers, provide good evidence for an interpretation of the integrated regulation of contractility in intact cardiac muscle (2,4).

CYCLIC AMP AND CARDIAC INOTROPIC RESPONSE

Synthesis of adenosine 3',5'-monophosphate (cyclic AMP) and the subsequent activation of protein kinase A (PKA) provide the most important facilitatory regulation of cardiac E-C coupling in situ (11). Elevations in levels of cyclic AMP in specific compartments in myocardial cells, due to selective activation of subtypes of β-adrenoceptors or selective inhibition of phosphodiesterase (PDE) isozymes have been a major focus of interest in relation to the therapeutic application of these agents (12,13).

Agents that Alter the Metabolism of Cyclic AMP

Accumulation of cyclic AMP in myocardial cells is achieved either by activation of adenylate cyclase or by inhibition of PDE (Fig. 1). Stimulation of β_1-receptors by

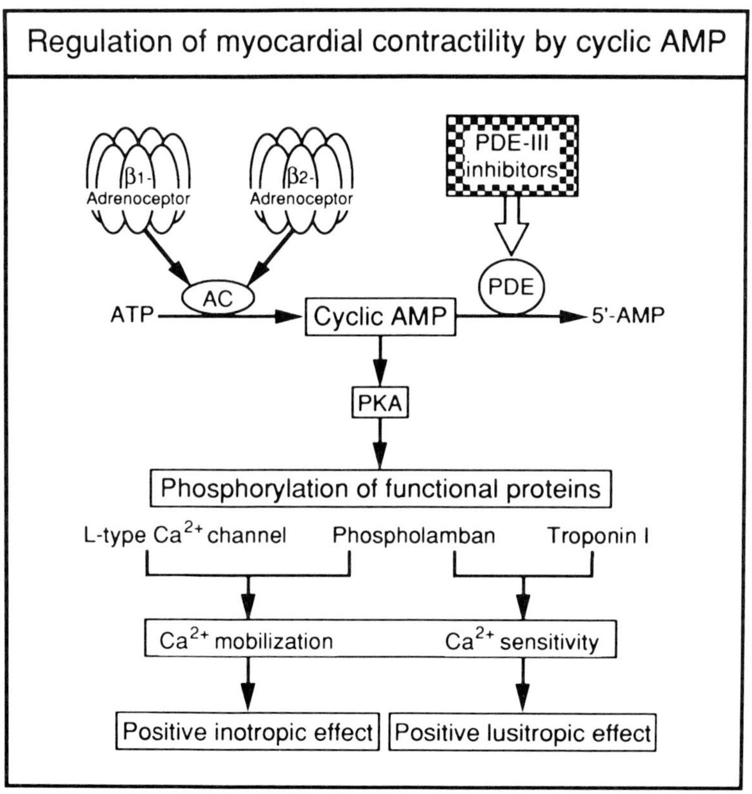

Fig. 1. Schematic representation of signal transduction pathway of cyclic AMP-mediated regulation of myocardial contractility.

sympathomimetic amines leads to the accumulation of cyclic AMP in myocardial tissues. Cyclic AMP is likewise generated in response to a number of exogenous agents that induce either the activation of adenylate cyclase, e.g., forskolin (14), glucagon (15), prostaglandins (16), cholera toxin (17), histamine (18), or the inhibition of PDE activity by PDE inhibitors such as 3-isobutyl-1-methylxanthine (IBMX), theophylline, amrinone, milrinone and enoximone (12,13). These agents, as well as lipophilic derivatives of cyclic AMP, such as dibutyryl cyclic AMP (db-cAMP), 8-thio-benzyl-cAMP and 8-bromo-cAMP (19) increase cardiac contractile function through common mechanisms, namely, by activation of cyclic AMP-dependent PKA, thereby promoting the phosphorylation of functional proteins (20) in myocardial cell membranes, the sarcoplasmic reticulum (SR) and myofibrils (Fig. 2). Injections of cyclic AMP and PKA into single myocardial cells (21) and photoactivation of an intracellularly applied photolabile analogue of cyclic AMP, the 4,5-dimethoxy-2-nitrobenzyl ester of cyclic AMP (22), mimic the functional and electrophysiological changes induced by the cardiotonic agents that act through accumulation of cyclic AMP.

Effects of Isoproterenol, a Full Agonist for β-Adrenoceptors

The Ca^{++}-sensitive bioluminescent protein aequorin (10) and fluorescent dyes, such as quin-2 (23), fura-2 (24) and indo-1 (25), have been successfully loaded into myocardial cells isolated from mammalian and amphibian hearts. In these intact myocardial cells, the changes in Ca^{++} transients during twitch contraction have been detected as light or fluorescent signals simultaneously with contractile force. In isolated cardiac muscle from the frog (11), cat (3), rat (38), rabbit (10), dog (14,27) and ferret (28,29), β-adrenoceptor agonists consistently induce a prominent increase in peak Ca^{++} transients and contractile force, in association with the abbreviation of both signals. The cyclic AMP-mediated changes in Ca^{++} transients and force are essentially similar for cardiac muscles isolated from a variety of mammalian species, but there is quite a wide range of quantitative variation in the mode of such regulation among species. For example, the apparent desensitization of myofibrils to Ca^{++} ions produced by β-adrenoceptor stimulation in the rabbit papillary muscle (8) was much less prominent than that observed in the cat papillary muscle (3,30). The difference might reflect species differences in the relative sensitivity to cyclic AMP-dependent phosphorylation of critical membrane proteins and contractile proteins, including troponin I, and troponin C.

The PIE of isoproterenol was associated with much higher increases in the amplitude of Ca^{++} transients (8). Therefore, for a given increase in force, strong β-adrenoceptor stimulation by isoproterenol often increases the amplitude of Ca^{++} transients to a level that exceeds the level achieved by an elevation of $[Ca^{++}]_o$. This result is consistent with the view that the cyclic AMP-dependent phosphorylation of troponin I leads to a decrease in myofibrillar sensitivity to Ca^{++} ions. However, a change in the affinity of troponin C for Ca^{++} ions is not considered to be the sole possible cause for a rightward deviation in the peak light-force curve induced by isoproterenol. The contractile apparatus may become saturated with Ca^{++} ions before the maximum effect of stimulation of β-receptors on Ca^{++} transients has been achieved, because the concentration-response curves for the increase in Ca^{++} transients induced by catecholamines, which are full agonists for β-adrenoceptors, are to the left of the concentration-response curves for the increase in force (8). In addition, the reduction in duration of Ca^{++} transients elicits the rightward shift of the relationship, because an equilibrium is not achieved during twitch contraction between the $[Ca^{++}]_i$ and force development (6). The decrease in myofibrillar responsiveness to Ca^{++} ions induced by β-adrenoceptor stimulation has been demonstrated unequivocally during tetanic contraction produced by high frequency electrical stimulation in the presence of ryanodine (26,29).

Effects of Denopamine, a Partial Agonist for β-Adrenoceptors

The possible cellular compartmentalization of cyclic AMP, as well as phosphodiesterase isozymes, (PKA and their substrates which are relevant to the regulation of myocardial function), has been postulated since cyclic AMP regulates multiple processes in myocardial cells (31-33). Tissue levels of cyclic AMP are often dissociated from the extent of functional modification in cardiac muscle.

Partial agonists of β_1-adrenoceptors, such as prenalterol (34), xamoterol (35) and denopamine (36), are less effective with respect to accumulation of cyclic AMP than the full agonist isoproterenol. However, the functional responses to these partial agonists are much greater than that expected from the accumulation of cyclic AMP, as compared with the effect of isoproterenol (37). One β_1-partial agonist, namely, prenalterol increases levels of free and bound cyclic AMP proportionally, whereas isoproterenol causes a greater increase in levels of free than in levels of bound cyclic AMP in the rat perfused heart (34). At functionally equal concentrations, prenalterol and isoproterenol increase levels of bound cyclic AMP to the same extent, while the increase in the level of free cyclic AMP is much greater with isoproterenol (34). These findings support the proposed compartmentalization of cyclic AMP in myocardial cells, and may indicate that β_1-partial agonists elevate levels of cyclic AMP in the functionally relevant compartment(s) more effectively than isoproterenol.

Although β_1-partial agonists cause the accumulation of much less cyclic AMP than isoproterenol for a given increase in force, it is unclear whether the nature of the regulation of Ca^{++} transients and of contractility by these agents is different from that by the nonselective full agonist isoproterenol. The partial β_1-agonist denopamine, with an intrinsic activity of 0.9 (the maximum PIE of isoproterenol $= 1.0$) was used to examine this issue in aequorin-injected ventricular muscle of the dog (38).

The bell-shaped concentration-response curve for the PIE of denopamine was associated with the bell-shaped curve for increase in Ca^{++} transients induced by denopamine. The relationship between peak force and the amplitude of Ca^{++} transients induced by denopamine was not different from that induced by isoproterenol. This equivalence indicates that, although denopamine, a partial agonist of β_1-adrenoceptors, causes the accumulation of cyclic AMP to a much lesser extent than the full agonist isoproterenol, the extent of accumulation of cyclic AMP in the functionally relevant compartment(s) may be identical and, therefore, the myofibrillar responsiveness to Ca^{++} ions is decreased by denopamine in a similar manner to the decrease caused by isoproterenol (38).

Effects of Milrinone and OPC 18790, Selective PDE Inhibitors

Phosphodiesterases may similarly be compartmentalized in myocardial cells because selective inhibition of PDE III bound to the SR, which may lead to localized increases in levels of cyclic AMP and activation of PKA, has been shown to be relevant to the PIE of PDE inhibitors in most mammalian cardiac muscles (39,40). It also appears that localization of the PDE III shows quite a wide range of variation among species of animals (40).

PDE inhibitors enhance and mimic the effects of β-adrenoceptor agonists on the electrical characteristics of membranes, Ca^{++} transients and contractile force. Newly developed cardiotonic agents, such as amrinone, milrinone, enoximone, piroximone, imazodan and LY 195115 selectively inhibit the PDE III isozyme which has a low Km for cyclic AMP and can be inhibited by cyclic GMP (41,42) bound to the cardiac SR membrane. Selective binding to and inhibition of this isozyme has been shown to be relevant to the PIE of these agents (39,43,44). The PIE of milrinone, amrinone, enoximone and piroximone was accompanied by increases in the peak Ca^{++} transient in association with increases in the rate of uptake of Ca^{++} ions by the SR (14). Accumulation of cyclic AMP, and the increase in peak Ca^{++} transients and force induced by these agents in the dog ventricular muscle can be abolished by carbachol, an indication that their effects are mediated by a cyclic AMP-dependent mechanism (14).

The regulation of Ca^{++} transients and contractility by selective inhibitors of PDE III was examined in a comparison with those induced by isoproterenol in the aequorin-injected ventricular muscle of the dog. The curves representing the relationship between peak Ca^{++} transients and force during induction of the PIE by two selective inhibitors of PDE III, namely, milrinone (Fig. 2) and OPC-18790 (Fig. 3), were superimposable on the curve obtained when $[Ca^{++}]_o$ was elevated, a result that indicates that the inhibitors of PDE III did not cause a decrease in myofibrillar responsiveness to Ca^{++} ions. The effects of these inhibitors differed from those of denopamine and isoproterenol (38). These findings support the proposed existence of different intracellular compartments of cyclic AMP, in which the levels of cyclic AMP may be regulated differentially by facilitation of the generation of cyclic AMP or by selective inhibition of its breakdown (31,33).

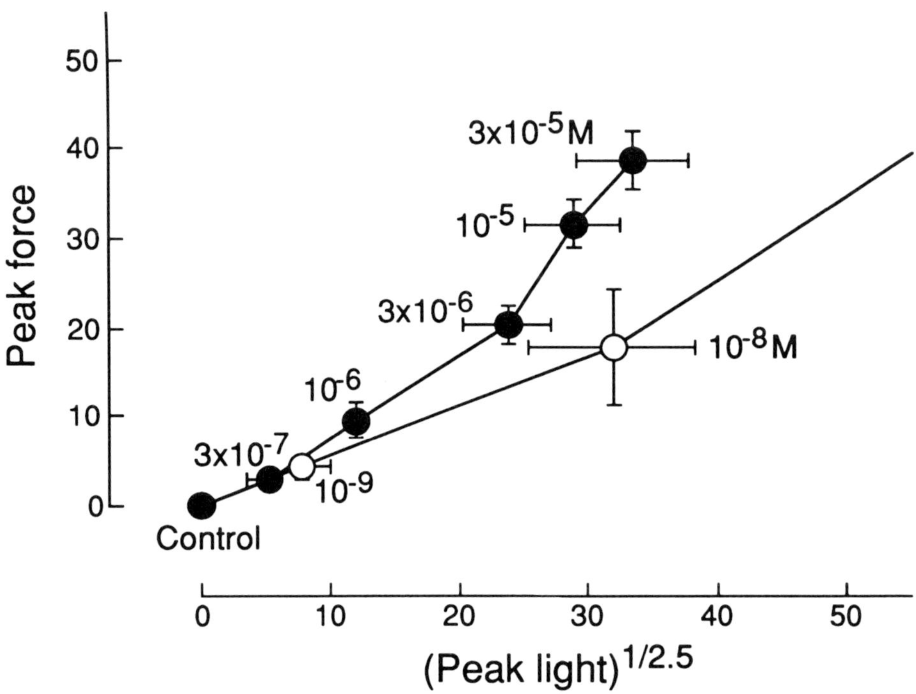

Fig. 2. The relation between the amplitude of aequorin light transients and that of isometric contractions as the contractility was varied by OPC-18790 or by isoproterenol in the same dog ventricular trabecula microinjected with aequorin. The 2.5th root of the amplitude of the aequorin light transients has been used as an indicator of the amplitude of Ca^{++} transients. Numbers adjacent to the curves indicate the molar concentration. Closed circles: OPC-18790; open circles: isoproterenol. Vertical and horizontal bars: $\pm$ SEM.

Rapundalo and coworkers (50) compared the effects of various concentrations of isoproterenol on force, levels of cyclic AMP and on the phosphorylation of phospholamban, troponin I, and troponin C with the effects on these preparations of milrinone and IBMX in the perfused guinea-pig heart. They showed that PDE inhibitors enhance contraction and relaxation in association with little or no concomitant change in phosphorylation of the SR and contractile proteins, despite relatively large increases in tissue levels of cyclic AMP. The results shown in Figs. 2 and 3 are consistent with these biochemical findings of Rapundalo et al. (50).

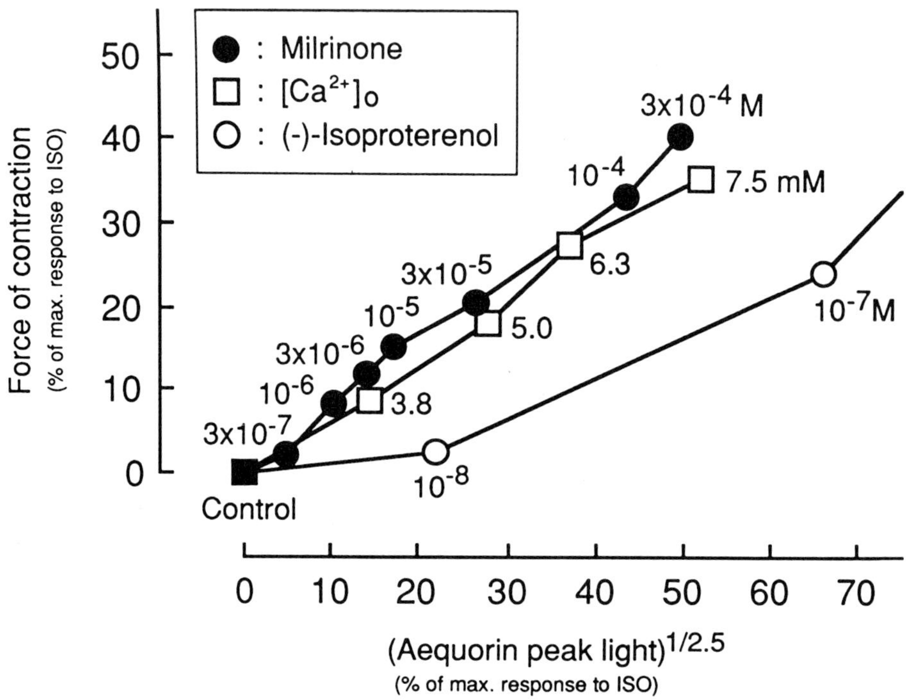

Fig. 3. The relation between the amplitude of aequorin light transients and that of isometric contractions as the contractility was varied by milrinone , alterations of $[Ca^{++}]_o$ or isoproterenol in a dog ventricular trabecula microinjected with aequorin. Numbers adjacent to the curves indicate the molar or millimolar concentration. Basal force and aequorin light: 6.21 mN/mm^2 and 2.76 nA. Maximum force and light determined by administration of isoproterenol: 26.6 mN/mm^2 and 41.9 nM, respectively.

THE HYDROLYSIS OF PHOSPHOINOSITIDE AND CARDIAC INOTROPIC RESPONSE

Increasing interest has also been focused on another pathway that involves the receptor-mediated acceleration of the hydrolysis of phosphoinositide (PI), and the resultant production of inositol 1,4,5-trisphosphate (IP_3), as well as diacylglycerol (DAG) that activates PKC. The role of this process is, however, less understood in myocardial cells. The characteristics of this pathway are coupling to multiple regulatory processes, which include various types of Ca^{++}, K^+ and Cl^- channels, ion-transport systems, such as Na^+-H^+ exchange, and responsiveness to Ca^{++} ions of contractile proteins (for reviews see 46-48). This transduction process shows a wide range of variation among different mammalian species. A series of endogenous transmitters and hormones, including catecholamines (via α_1-adrenoceptor), muscarinic receptor agonists (via M_1- or M_3-receptor), angiotensin II (via AT_1-ATP receptor), (via P_2-receptor), vasopressin (via V_1-receptor) and histamine (H_1 receptor), stimulate this process.

Agonists of α_1-Adrenoceptors

The positive inotropic effects of sympathomimetic amines are partially mediated by α_1-adrenoceptors in most mammalian cardiac muscles (49), and stimulation of myocardial α_1-adrenoceptors does accelerate the hydrolysis of PI (50). In the mammalian myocardium, most sympathomimetic amines, including norepinephrine, epinephrine, dopamine and phenylephrine, modulate both the Ca^{++} transient and contractility via activation of α_1-adrenoceptors (46-48). While α-stimulation increases both the amplitude of Ca^{++} transients and the force of contraction in aequorin-injected papillary muscles of the rabbit, the increase in Ca^{++} transients is much smaller than that of alteration of $[Ca^{++}]_o$ (8). The maximum increase in Ca^{++} transients was less than 10% of the maximum response to isoproterenol. The maximal increase in force induced by α-stimulation is about 50-60% of that induced by isoproterenol (8). Hence, for a given increase in the force of contraction, stimulation of α-adrenoceptors produces much smaller change in the amplitude of Ca^{++} transients than does elevation of $[Ca^{++}]_o$ or β-adrenoceptor stimulation, an indication that the PIE of α-stimulation may, in large part, be the result of an increase in myofibrillar responsiveness to Ca^{++} ions (8). In support of this suggestion, stimulation of α-adrenoceptors reduces the duration of Ca^{++} transients in association with a prolongation of isometric contractions (8). The activation of myocardial α_1-adrenoceptors may, therefore, produce a PIE through both an increase in mobilization of intracellular Ca^{++} ions and an increase in myofibrillar responsiveness to Ca^{++} ions.

Experimental evidence indicates that multiple processes, triggered by the stimulation of α_1-adrenoceptors are responsible for increased mobilization of Ca^{++} ions and regulation of the myofibrillar responsiveness to Ca^{++} ions. These processes include an increase in the influx of Ca^{++} ions through L-type Ca^{++} channels and the activation of Na^+-H^+ exchange and subsequent modulation of Na^+-Ca^{++} exchange (51,53). The activation of L-type Ca^{++} channels is not directly facilitated by activation of α_1-adrenoceptors in whole-cell voltage-clamped cardiomyocytes and, therefore, this activity may be increased indirectly via inhibition of K^+ channels and resultant prolongation of the duration of action potentials (54-56). It has been shown that currents through cardiac K^+ channels, including the transient outward current, I_{to}, in atrial and ventricular (rat) myocytes and the inward rectifier K^+ current, I_{k1}, in rabbit ventricular myocytes, are suppressed by the activation of α_1-adrenoceptors (57,58). The organic Ca^{++} antagonists (nifedipine, verapamil and diltiazem) effectively inhibited the α_1-mediated PIE to an extent similar to the extent of inhibition of the basal force of contraction (59), but the extent of the contributions of L-type Ca^{++} channels, Na^+-K^+ exchange and increased myofibrillar responsiveness to Ca^{++} ions to the α_1-mediated PIE remain to be determined.

The possible involvement of two processes has been postulated as the mechanism of the α_1-adrenoceptor-mediated increase in myofibrillar responsiveness to Ca^{++} ions: 1) an intracellular alkalinization induced by activation of the Na^+-H^+ exchanger; and 2) the phosphorylation of contractile proteins. It has been suggested that the contribution of the former mechanism may be less significant in the rat because an α_1-mediated increase in myofibrillar responsiveness to Ca^{++} ions is elicited even in the presence of an inhibitor of the Na^+-H^+ exchange, namely, ethylisopropylamiloride (EIPA) at 0.1 μM (53). However, Gambassi and coworkers (60) showed that EIPA at 10 μM abolished the alkalinization and the increase in myofibrillar responsiveness to Ca^{++} ions induced via α-adrenoceptors in ventricular myocytes of the rat. In isolated, paced left atria of the rat, another selective inhibitor of Na^+-H^+ exchange, hexamethylamiloride (10 μM) prevented 50% of the PIE of phenylephrine (mediated via α-adrenoceptors)(61). Since the selectivity of these compounds on the Na^+-H^+ exchange is not very high, further studies with more selective agents may be necessary before we can draw any conclusion about the significance of alkalinization.

It was shown that PKC phosphorylates troponin I and troponin T *in vitro* (62). However, Edes and Kranias (63) failed to detect the phosphorylation of troponin in the perfused guinea-pig heart. It has been proposed that such phosphorylation may be related to the negative inotropic effect induced by externally applied phorbol esters in cardiac muscles from various mammals (64-67), but in the cardiac muscles of certain species a low concentration of phorbol esters exert a PIE (68-70). Clearly, the physiological significance of the PKC-induced phosphorylation of contractile proteins requires further study.

In the rabbit papillary muscle, phorbol esters, namely, PDBu (70,71) and 12-O-tetradecanoyl phorbol-13-acetate (72) inhibit the PIE mediated via α_1-adrenoceptors. At the concentration that inhibits the α_1-mediated PIE, PDBu increases the basal force and does not attenuate the PIE mediated by -adrenoceptors, an indication that the inhibitory effect of PDBu is selective. While the binding characteristics of α_1-adrenoceptors are unaffected by PDBu, the acceleration of the hydrolysis of PI induced by α_1-stimulation is decreased, indicating that PDBu may cause an uncoupling of the receptor to activation of phospholipase C, probably acting on the GTP binding protein, Gq, or directly on the phospholipase C (70). By contrast, in the rat papillary muscle, PDBu enhanced the α_1-mediated PIE (72), an indication that the effects of phorbol esters exhibit considerable variation among species. Inhibitors of PKC, namely, staurosporine, NA 0345 and H-7, also selectively inhibit the α_1-adrenoceptor-mediated PIE in the rabbit ventricle (73).

Activation of myocardial α_1-adrenoceptors in certain cardiac tissues is also coupled to cellular processes that may lead to a decrease in $[Ca^{++}]_i$ or that counteract an increase in $[Ca^{++}]_i$. For example, in canine Purkinje fibers, α_1-stimulation activates the Na^+-K^+ ATPase via the PTX-sensitive G protein. This activation may reduce $[Na^+]_i$ and thereby lower $[Ca^{++}]_i$(74). In guinea-pig ventricular myocytes, the activation of PKC by phorbol esters and/or the stimulation of α_1-adrenoceptors activates the delayed rectifier K^+ current, I_K (75), and the K^+ current activated via muscarinic ACh receptors, $I_{k,ACh}$ (76), which may also lead to a decrease in $[Ca^{++}]_i$. These divergent signal-transduction processes, triggered by activation of myocardial α_1-adrenoceptors, may contribute to a wide range of species-dependent variations in the α_1-mediated PIE (49). Species-dependent variations in the distribution of α_1-adrenoceptors may also contribute to the species-dependent variations in functional regulation (77).

Endothelin

Endothelin (ET) is an endogenous vasoconstrictor polypeptide of 21 amino acids which was first isolated from porcine aortic endothelial cells (78) an abundant receptors for ET have also been found in myocardial cell membranes (79,80). ET has a pronounced PIE on cardiac muscles of mammals that include the rat, guinea pig, ferret, rabbit and human (80-82). The PIE of ET closely resembles that of α_1-adrenoceptor agonists in a number of respects. Although the extent of the PIE coincides approximately with the density of receptors for ET in various mammalian species, ET does not have a PIE on the dog ventricle in spite of the presence of receptors for ET (80). This result is similar to the findings with α_1-adrenoceptor agonists in the dog ventricle (77,83).

The PIE of ET is accompanied by acceleration of the hydrolysis of PI in cardiac tissue (84). The dependence on time and concentration of the PIE and the hydrolysis of PI are very well correlated in the rabbit ventricular myocardium (85). The PIE of ET is associated with prolongation of the total duration of contraction and retardation of relaxation (a negative lusitropic effect) (80), which are likewise characteristic of α_1-adrenoceptor-mediated modulation of isometric contractions.

In rat ventricular myocytes, ET has a PIE in association with intercellular alkalinization that is due to activation of the Na^+-H^+ exchanger (86). ET also affects activities of the K^+ and Ca^{++} channels. Habuchi and coworkers (87) showed that, in guinea-pig ventricular myocytes, ET-1 enhances the I_K current without affecting the L-type Ca^{++} current, as do α_1-adrenoceptor agonists. By contrast, the activity of rat cardiac K^+ channels (carrying I_k) expressed in *Xenopus* oocytes is suppressed by ET-1 via co-expressed receptors for ET in a process that probably involves the hydrolysis of PI (88). In spontaneously contracting atrial cells of the rat, all three isopeptides of ET decrease the rate of beating and cause an increase in I_{K1} current with ET-3 being the most potent of the three (89). Tohse and coworkers (90) reported that, in guinea-pig ventricular myocytes, ET decreases the amplitude of I_{ca} measured under whole-cell voltage-clamp conditions. By contrast, in rabbit cardiomyocytes, when GTP (100 μM) is added to the pipette solution used to dialyze the inside of the cell, ET-1 causes a large, reproducible increase in peak I_{ca} via a PTX-insensitive process (91). Thus, available information about the ET-mediated regulation of ion channels is still fragmentary and this topic requires further study. It has been demonstrated in the ferret papillary muscle loaded with aequorin that ET-1 causes an increase in the amplitude of Ca^{++} transients in association with an increase in responsiveness of myofibrils to Ca^{++} ions (92). These findings are very similar to those obtained previously with α_1-adrenoceptor agonists that also accelerate the hydrolysis of PI in the rabbit papillary muscle (77).

In the rabbit ventricular muscle, the α_1-adrenoceptor-mediated PIE is selectively inhibited by tumor-promoting phorbol esters. PDBu similarly antagonizes the PIE of ET-1 concomitant with inhibition of the ET-induced hydrolysis of PI at concentrations at which PDBu does not affect the PIE of isoproterenol or of Bay k 8644 (80).

Cloning of a cDNA for the receptor for ET demonstrated the existence of at least two subtypes, namely, ET_A and ET_B, in vascular smooth muscle and endothelial cells. The ET_A receptor is relatively selective for ET-1, whereas the ET_B receptor is nonselectively activated by ET-1 and ET-3 (93). An additional receptor, designated ET_C, that is preferentially activated by ET-3 was discovered more recently (93). In the rabbit papillary muscle, ET-1, ET-2 and ET-3 produce a PIE with approximately the same potency and efficacy (80). A detailed examination of the concentration-response curve for ET isopeptides has shown, however, that ET-1 has a biphasic PIE: there is an initial phase at 10^{-14}-10^{-10} M, a range at which ET-3 is ineffective; and a second phase over a range of concentrations above 10^{-10} M, which can be elicited by either ET-1 or ET-3, whose actions during this phase are equivalent (94). Pharmacological analysis using newly developed subtype-selective antagonists or agonists, such as BQ-123 or IRL 1620, and two derivatives of sarafotoxin, namely, [Thr^2]SRTB and [Glu^9]SRTB, has shown that the receptors for ET in the rabbit ventricular myocardium are neither the conventional ET_A nor ET_B receptors (94).

Angiotensin II

The systemic and local renin-angiotensin systems play an important role in the control of cardiovascular hemodynamics and pathophysiologic regulation via activation of receptors for angiotensin II (Ang II). In addition to its well documented vasoconstrictor effect, Ang II has specific cardiac effects, such as facilitation of cardiac hypertrophy and fibrosis (95), as well as positive chronotropic and inotropic effects in avian hearts (96,97) and in mammalian hearts including those of the cat, rabbit, calf (98), neonatal rat (99,100) and human (101). The induction of a PIE by Ang II appears to be markedly dependent on a variety of modulatory factors, which include different regions (atrial or ventricular), species, and the presence or absence of endothelium (102). For example, whereas in preparations of isolated dog and adult rat heart Ang II does not have a PIE, it has positive inotropic and chronotropic effects, mediated by the direct activation of cardiac receptors for Ang II, in the isolated blood-perfused dog heart preparation (103) and the pithed rat heart (104). It has been reported that, in guinea-pig atrial and ventricular muscles, Ang II does not have any PIE (96,105). However, recently, Ang II was shown to have a PIE in guinea-pig atria (106). The rabbit cardiac muscle contains receptors with high affinity for Ang II (107,108) and it responds consistently to Ang II with a pronounced PIE (96). The cellular mechanism of the Ang II-induced PIE has been postulated to involve facilitation of L-type Ca^{++} channels because the slow inward current (I_{si}) is increased by Ang II (96,98). It has also been shown that Ang II accelerates the hydrolysis of PI in cardiac muscle but a role of this process in cardiac regulation has not yet been established. For example, while the hydrolysis of PI is clearly accelerated by Ang II in guinea-pig cardiac muscle, a PIE is absent (105). By contrast, in the avian heart, Ang II induces both a PIE and the hydrolysis of PI (97). In the rabbit ventricular muscle, the Ang II-induced PIE and hydrolysis of PI are well correlated in terms of dependence on time and concentration (109). Moreover, saralasin and a selective antagonist of AT_1-receptors, losartan, but not a selective antagonist of AT_2-receptors, PD 123319, antagonize the Ang II-induced PIE and hydrolysis of PI (109). The PIE of Ang II closely resembles that of α_1-stimulation and ET: 1) the PIE is associated with a characteristic prolongation of isometric contractions; 2) PDBu (10-100 nM) selectively antagonizes the PIE and the hydrolysis of PI induced by Ang II, an observation that suggests that the hydrolysis of PI is a common mediator of the PIE induced by activation of this class of receptor. While Ang II facilitates the opening of membrane Cl^- channels in rabbit ventricular myocytes (110), the functional relevance of such regulation in the inotropic response remains to be established.

CA^{++} SENSITIZERS

Cardiotonic agents, such as EMD 53998 and Org 30029, which act via a novel mechanism to increase myofibrillar responsiveness to Ca^{++} ions, have received more attention in terms of potential usefulness in clinical application (12,13). These agents either affect the central mechanism or act via the downstream mechanism. Physical interventions that modulate cardiac contractility through subcellular mechanisms that are independent of cyclic nucleotides and the hydrolysis of PI include the effects of alterations in the frequency of contraction, muscle length, temperature, and $[Ca^{++}]_o$. Alterations in $[Ca^{++}]_o$ affect contractility essentially via changes in Ca^{++} transients. Since the binding of Ca^{++} ions to regulatory sites on troponin C is length-dependent (111), changes in muscle length modify myofibrillar responsiveness to Ca^{++} ions (112-113).

Ca^{++} sensitizers, such as sulmazole (14), EMD 53998 (4) Org 30029 (114), MCI 154 (115) and pimobendan (116), are cardiotonics with novel mechanism(s) of action. For example, a significant portion of the contractile response to sulmazole is considered to result from an increase in myofibrillar responsiveness to Ca^{++} ions, as observed also in *in vitro* studies (117). In the dog ventricular muscle, elevation of levels of cyclic AMP after administration of 0.2 mM sulmazole becomes significant only after the PIE has reached a steady level; a result that contrasts strongly to the similar time course of increases in levels of cyclic AMP and force after the administration of enoximone (118). In aequorin-injected ventricular trabeculae of the dog, the increase in peak Ca^{++} transients, as well as the PIE, and the accumulation of cyclic AMP produced by enoximone are abolished by carbachol, while the effects of sulmazole on the Ca^{++} transient and its interaction with carbachol are more complicated (118). Since the effects of high concentrations of sulmazole are resistant to carbachol, it appears that sulmazole elicits its PIE via a cyclic AMP-dependent mechanism in lower concentrations and via an increase in the responsiveness of myofilaments to Ca^{++} ions at higher concentrations. The reduction in peak Ca^{++} transients elicited by higher concentrations of sulmazole may be due in part to inhibition of the uptake of Ca^{++} ions by the SR, as is also observed in the case of other inhibitors of PDE, such as caffeine and theophylline. As a whole, these effects of caffeine and theophylline in preparations of intact myocardium are consistent with those in skinned or hyperpermeable cardiac muscle. Caffeine and theophylline each increase the myofibrillar responsiveness to Ca^{++} ions in mammalian ventricular fibers including those of the guinea pig (119) and rat (120). They are currently receiving attention in terms of possible clinical application in the treatment of patients with heart failure.

Newly developed Ca^{++}-sensitizers are attractive as potential therapeutic agents for the treatment of patients with congestive heart failure because 1) the activation energy required for the transport of Ca^{++} ions may be preserved with this class of agents; 2) Ca^{++} overload, which may be responsible for the arrhythmogenic actions of digitalis and catecholamines and the ultimate death of myocardial cells, can be avoided; and 3) these agents provide a potential novel mechanism for reversing the myocardial contractile dysfunction encountered in ischemia and reperfusion.

In the aequorin-injected ventricular trabeculae of the dog, Org 30029 increased the isometric contraction to a level higher than the maximum response to isoproterenol and the PIE of Org 30029 was associated with much less increase in Ca^{++} transients than with the inotropic response to isoproterenol (114). Among Ca^{++}-sensitizers, the effect of a derivative of diazinone designated EMD 53998 has been studied most extensively. EMD 53998 improves the economy of chemo-mechanical energy transduction in cardiac muscle (121). In different studies, the PIE of EMD 53998 has been associated with a variable effect on the Ca^{++} transient ranging from no change to a small increase or decrease (122,123). Available preparations of EMD 53998 consist of a racemic mixture with both PDE-inhibitory action and Ca^{++}-sensitizing action. Recently, it was shown that the (+)-isomer has a predominantly Ca^{++}-sensitizing action, while the (-)-isomer is the inhibitor of PDE (124). These findings are important and intriguing because a number of classical and new agents, such as caffeine, theophylline, sulmazole, pimobendan, MCI-154 and Org 30029 (114,116,1-25), have both PDE-inhibitory and Ca^{++}-sensitizing actions. The selective actions of the enantiomers of EMD 53998 suggest that the binding sites or sites of action on PDE III and contractile proteins may be closely related in terms of structure. EMD 53998 or MCI-154 can reverse the contractile depression induced by anoxia (122), acidosis (126) and inorganic phosphate (127), actions which may have a critical modulating effect on cardiac contractility under ischemic conditions.

Although the mechanism of drug-induced myofibrillar sensitization to Ca^{++} ions has not been unequivocally elucidated, it appears that, in addition to modulating the affinity of troponin C for Ca^{++} ions, EMD 53998 may exhibit an actin-myosin interface as its potential site of action and cause a change in cycling of crossbridges (4). EMD 53998 activates actomyosin ATPase in the absence of Ca^{++}-regulatory proteins and increases the rate of interaction of actin with myosin in a motility assay in the absence of Ca^{++} ions (4). The latter effect does not involve Ca^{++} ions or is operative at a site beyond Ca^{++}-signaling events and, therefore, the term "Ca^{++}-sensitizer" is no longer applicable.

CONCLUSION

The signal transduction pathway mediated by cyclic AMP that leads to the characteristic PIE in association with a positive lusitropic effect (acceleration of rate of twitch relaxation) has been well established. Relationships between accumulation of cyclic AMP, changes in intracellular Ca^{++} transients and the PIE differ, however, depending on the mechanism of particular drugs that affect different steps in the metabolism of cyclic AMP. Partial agonists of β_1-adrenoceptors and inhibitors of PDE III cause the accumulation of less cyclic AMP for a given extent of PIE than does isoproterenol. In addition, in aequorin-microinjected canine ventricular muscle, selective inhibitors of PDE III, milrinone and OPC 18790, scarcely produced a shift to the right of the relationship between the amplitude of force vs. Ca^{++} transients, while a partial agonist of β_1-adrenoceptors, denopamine, elicits a decrease in Ca^{++} responsiveness to the same extent as isoproterenol.

Activation of myocardial α_1-adrenoceptors, as well as stimulation of receptors for endothelin and angiotensin II, which accelerate hydrolysis of PI to result in production of IP_3 and DAG are associated with very similar inotropic regulation: 1) the dependence on the species of animals of induction of the PIE; 2) an excellent correlation between the extent of acceleration of hydrolysis of PI and the PIE; 3) isometric contraction curves associated with a negative lusitropic effect; 4) the PIE associated with increases in myofibrillar responsiveness to Ca^{++} ions; and 5) the selective inhibition of the PIE by an activator of PKC, PDBu, with little effect on the PIE of isoproterenol and Bay k 8644.

A novel class of cardiotonic agents, namely, Ca^{++} sensitizers such as EMD 53998 and Org 30029, act on the Ca^{++}-binding site of troponin C, increasing the affinity of these sites for Ca^{++} ions, or at the actin-myosin interface to facilitate the cycling of crossbridges. These agents produce a PIE with little change or a decrease in Ca^{++} transients and may bring about a significant breakthrough in the development of drugs for the reversal of myocardial failure in the treatment of congestive heart failure.

ACKNOWLEDGEMENT

This work was supported in part by Grants-in-Aid for Scientific Research on Priority Area (no. 03253102), Scientific Research (B) (no. 03454142) and Developmental Scientific Research (B) (no. 04557009) from the Ministry of Education, Science and Culture, Japan.

REFERENCES

1. Blinks, J.R. and Endoh, M. Modification of myofibrillar responsiveness to Ca^{++} as an inotropic mechanism. Circulation 73 (suppl. III):III-85-III-98, 1986.

2. Robertson, S.P., Johnson, J.D., Holroyde, M.J., Kranias, E.G., Potter, J.D. and Solaro, R.J. The effect of troponin I phosphorylation on the Ca^{++}-binding properties of the Ca^{++}-regulatory site of bovine cardiac troponin. J. Biol. Chem. 257:260-263, 1982.

3. Allen, D.G. and Kurihara, S. Calcium transients in mammalian ventricular muscle. Eur. Heart J. 1(suppl. A):5-15, 1980.

4. Solaro, R.J., Gambassi, G., Warshaw, D.M., Keller, M.R., Spurgeon, H.A., Beier, N. and Lakatta, E.G. Stereoselective actions of thiadiazinones on canine cardiac myocytes and myofilaments. Circ. Res. 73:981-990, 1993.

5. Reiter, M. Calcium mobilization and cardiac inotropic mechanisms. Pharmacol. Rev. 40:189-21, 1988.

6. Yue, D.T. Intracellular [Ca^{++}] related to rate of force development in twitch contraction of heart. Am. J. Physiol. 252:H760-H770, 1987.

7. Blinks, J.R. Analysis of the effects of drugs on myofibrillar Ca^{++} sensitivity in intact cardiac muscle. In: Modulation of Cardiac Calcium Sensitivity: A New Approach to Increasing the Strength of the Heart, eds. Lee, J.A. and Allen, D.G. 242-282, 1993, Oxford University Press, Oxford.

8. Endoh, M., Blinks, J.R. Actions of sympathomimetic amines on the Ca^{++} transients and contractions of rabbit myocardium: reciprocal changes in myofibrillar responsiveness to Ca^{++} mediated through α- and -Adrenoceptors. Circ. Res. 62:247-265, 1988.

9. Allen, D.G. and Blinks, J.R. Calcium transients in aequorin-injected frog cardiac muscle. Nature 273:509-513, 1978.

10. Blinks, J.R., Wier, W.G., Hess, P. and Prendergast, F.G. Measurement of Ca^{++} concentrations in living cells. Prog. Biophys. Molec. Biol. 40:1-114, 1982.

11. Tsien, R.W. Cyclic AMP and contractile activity in heart. Adv. Cyclic Nucleo-tide Res. 8:363-420, 1977.

12. Farah, A.E., Alousi, A.A. and Schwarz, R.P., Jr. Positive inotropic agents. Ann. Rev. Pharmacol. Toxicol. 24:275-328, 1984.

13. Endoh, M. and Hori, M. Basic pharmacology and clinical application of new positive inotropic agents. Drugs of Today 29, 29-56, 1993.

14. Endoh, M., Yanagisawa, T., Taira, N. and Blinks, J.R. Effects of new inotropic agents on cyclic nucleotide metabolism and calcium transients in canine ventricular muscle. Circulation 73(suppl. III):III-117-III-133, 1986.

15. England, P.J. Studies on the phosphorylation of the inhibitory subunit of troponin during modification of contraction in perfused rat heart. Biochem. J. 160:295-304, 1976.

16. Brunton, L.L., Hayes, J.S. and Mayer, S.E. Hormonally specific phosphorylation of cardiac troponin I and activation of glycogen phosphorylase. Nature 280:78-80, 1979.

17. Endoh, M., Maruyama, M. and Yanagisawa, T. Cholera toxin-induced changes in force of contraction and cyclic AMP levels in canine ventricular myocardium: inhibition by carbachol. Life Sci. 35:2397-2406, 1984.

18. Levi, R., Allan, G., and Zavecz, J.H. Cardiac histamine receptors. Fed. Proc. 35:1942-1947, 1976.

19. Muneyama, K., Bauer, R.J., Shuman, D.A., Robins, R.K. and Simon, L.N. Chemical synthesis and biological activity of 8-substituted adenosine 3',5'-cyclic monophosphate derivatives. Biochemistry 10:2390-2395. 1971.

20. Solaro, R.J. Protein Phosphorylation in Heart Muscle CRC Press, Boca Raton, Florida, 1986.

21. Osterrieder, W., Brum, G., Hescheler, J., Trautwein, W., Flockerzi and V., Hofmann, F. Injection of subunits of cyclic AMPαdependent protein kinase into cardiac myocytes modulates Ca^{++} current. Nature 298:576-578, 1982.

22. Nerbonne, J.M., Richard, S., Nargeot, J. and Lester, H.A. New photoactivatable cyclic nucleotides produce intracellular jumps in cyclic AMP and cyclic GMP concentrations. Nature 310:74-76, 1984.

23. Lattanzio, F.A., Jr. and Pressman, B.C. Alterations in intracellular calcium activity and contractility of isolated perfused rabbit hearts by ionophores and adrenergic agents. Biochem. Biophys. Res. Commun. 139:816-821, 1986.

24. Cannell, M.B., Berlin, J.R. and Lederer, W.J. Effect of membrane potential changes on the calcium transient in single rat cardiac muscle cells. Science 238:1419-1423,1987.

25. Spurgeon, H.A., Stern, M.D., Baartz, G., Raffaeli, S., Hansford, R.G., Talo, A., Lakatta, E.G. and Capogrossi, M.C. Simultaneous measurement of Ca^{++}, contraction and potential in cardiac myocytes. Am. J. Physiol. 258:H574-H586, 1990.

26. Kurihara, S. and Konishi, M. Effects of -adrenoceptor stimulation on intracellular Ca transients and tension in rat ventricular muscle. Pfl|gers Arch. 409:427-437, 1987.

27. Endoh, M. Regulation of myocardial contractility via adrenoceptors: differential mechanisms of α-and -adrenoceptor-mediated actions. In: New Aspects of the Role of Adrenoceptors in the Cardiovascular System, eds. Grobecker, H., Philippu, A. and Starke, K. 78-105, 1986, Springer-Verlag Berlin Heidelberg.

28. McIvor, M.E., Orchard, C.H. and Lakatta, E.G. Dissociation of changes in apparent myofibrillar Ca^{++} sensitivity and twitch relaxation induced by adrenergic and cholinergic stimulation in isolated ferret cardiac muscle. J. Gen. Physiol. 92:509-529, 1988.

29. Okazaki, O., Suda, N., Hongo, K., Konishi, M. and Kurihara, S. Modulation of Ca^{++} transients and contractile properties by -adrenoceptor stimulation in ferret ventricular muscles. J. Physiol., 423:221-240, 1990.

30. Morgan, J.P. The effects of digitalis on intracellular calcium transients in mammalian working myocardium as detected with aequorin. J. Mol. Cell. Cardiol. 17:1065-1075, 1985.

31. Corbin, J.D., Sugden, P.H., Lincoln, T.M. and Keely, S.L. Compartmentalization of adenosine 3':5'-monophosphate and adenosine 3':5'-monophosphate-dependent protein kinase in heart tissue. J. Biol. Chem. 252:3854-3861, 1977.

32. Earp, H.S. and Steiner, A.L. Compartmentalization of cyclic nucleotide-mediated hormone action. Ann. Rev. Pharmacol. Toxicol. 18:431-459, 1978.

33. Buxton, I.L.O. and Brunton, L.L. Compartments of cyclic AMP and protein kinase in mammalian cardiomyocytes. J. Biol. Chem. 258, 10233-10239, 1983.

34. Aass H., Skomedal T. and Osnes, J.-B. Increase of cyclic AMP in subcellular fractions of rat heart muscle after -adrenergic stimulation: prenalterol and isoprenaline caused different distribution of bound cyclic AMP. J. Mol. Cell. Cardiol. 20:847-860, 1988.

35. Main, B.G. Structure-activity relations of -adrenergic agents. J. Chem. Tech. Biotechnol. 32:617-626, 1982.

36. Inamasu, M., Totsuka, T., Ikeo, T., Nagao, T. and Takayama, S. Beta1-adren ergic selectivity of the new cardiotonic agent denopamine in its stimulating effects on adenylate cyclase. Biochem. Pharmacol. 36:1947-1954, 1987.

37. Sasaki, Y., Yabana, H., Nagao, T. and Takayama, S. Effect of denopamine on the phosphorylation of cardiac muscle proteins in the perfused guinea-pig heart. Biochem. Pharmacol. 37:679-686, 1988.

38. Kohi, M., Norota, I., Takanashi, M. and Endoh, M. On the mechanism of action of the beta-1 partial agonist denopamine in regulation of myocardial contractility: effects on myocardial alpha adrenoceptors and intracellular Ca^{++} transients. J. Pharmacol. Exp. Ther. 265:1292-1300. 1993.

39. Kauffman, R.F., Crowe, V.G., Utterback, B.G. and Robertson, D.W. LY1951 15: A potent, selective inhibitor of cyclic nucleotide phosphodiesterase located in the sarcoplasmic reticulum. Mol. Pharmacol. 30:609-616,1986.

40. Weishaar, R.E., Kobylarz-Singer, D.C., Steffen, R.P. and Kaplan, H.R. Sub-classes of cyclic AMP-specific phosphodiesterase in left ventricular muscle and their involvement in regulating myocardial contractility. Circ. Res. 61: 539-547, 1987.

41. Harrison, S.A., Reifsnyder, D.H., Gallis, B., Cadd, G.G. and Beavo, J.A. Isolation and characterization of bovine cardiac muscle cGMP-inhibited phosphodiesterase: a receptor for new cardiotonic drugs. Mol. Pharmacol. 29: 506-514, 1986.

42. Weishaar, R.E., Burrows, S.D., Kobylarz, D.C., Quade, M.M. and Evans,D.B. Multiple molecular forms of cyclic nucleotide phosphodiesterase in cardiac and smooth muscle and in platelets. Isolation, characterization, and effects of various reference phosphodiesterase inhibitors and cardiotonic agents. Biochem. Pharmacol. 35:787-800, 1986.

43. Kauffman, R.F., Utterback, B.G. and Robertson, D.W. Specific binding of [3H]LY 186126, an analogue of indolidan (LY195115), to cardiac membranes enriched in sarcoplasmic reticulum vesicles. Circ. Res. 64:1037-1040, 1989.

44. Kithas, P.A., Artman, M., Thompson, W.J. and Strada, S.J. Subcellular distribution of high-affinity type IV cyclic AMP phosphodiesterase activity in rabbit ventricular myocardium: relations to the effects of cardiotonic drugs. Circ. Res. 62:782-789, 1988.

45. Rapundalo, S.T., Solaro, R.J. and Kranias, E.G. Inotropic responses to isoproterenol and phosphodiesterase inhibitors in intact guinea pig heartS: comparison of cyclic AMP levels and phosphorylation of sarcoplasmic reticulum and myofibrillar proteins. Circ. Res. 64:104-111, 1989.

46. Endoh, M. Signal transduction of myocardial α1-adrenoceptors: regulation of ion channels, intracellular calcium, and force of contraction - a review. J. Appl. Cardiol. 6:379-399, 1991.

47. Fedida, D., Braun, A.P. and Giles, W.R. α1-Adrenoceptors in Myocardium: functional aspects and transmembrane signaling mechanisms. Physiol. Rev. 73: 469-487, 1993.

48. Terzic, A., Puceat, M., Vassort, G. and Vogel, S.M. Cardiac α1-adrenocep tors: an overview. Pharmacol. Rev. 4: 147-175, 1993.

49. Endoh, M. Adrenoceptors and the myocardial inotropic response: do alpha and beta receptor sites functionally coexist? In: Trends in autonomic pharmacology, ed. Kalsner, S. vol 2:303-322, Urban & Schwarzenberg, Baltimore Munich, 1982.

50. Brown, J.H. and Martinson, E.A. Phosphoinositide-generated second messengers in cardiac signal transduction. Trends Cardiovasc. Med. 2:209-214, 1992.

51. Iwakura, K., Hori, M., Watanabe, Y. Kitabatake, A., Cragoe, E.J., Jr., Yoshida, H. and Kamada, T. α1-Adrenoceptor stimulation increases intracellular pH and Ca^{++} in cardiomyocytes through Na+/H+ and Na+/Ca^{++} exchange. Eur. J. Pharmacol. 186:29-40, 1990.

52. Puceat, M., Clement, O., Lechene, P., Pelosin, J.M., Ventura-Clapier, R. and Vassort, G. Neurohormonal control of calcium sensitivity of myofilaments in rat single cells. Circ. Res. 67:517-524,1990.

53. Terzic, A., Puceat, M., Clement, O., Scamps, F. and Vassort, G. α1-Adrenergic effects on intracellular pH and calcium and on myofilaments in single rat cardiac cells. J. Physiol. 447:275-292, 1992.

54. Hartmann, H.A., Mazzocca, N.J., Kleiman, R.B. and Houser, S.R. Effects of phenylephrine on calcium current and contractility of feline ventricular myocytes. Am. J. Physiol. 255:H1173-H1180, 1988.

55. Hescheler, J., Nawrath, H., Tang, M. Trautwein, W. Adrenoceptor-mediated changes of excitation and contraction in ventricular heart muscle from guinea-pigs and rabbits. J. Physiol. 397:657-670. 1988.

56. Fedida, D. and Bouchard, R.A. Mechanisms for the positive inotropic effect of α1-adrenoceptor stimulation in rat cardiac myocytes. Circ. Res. 71:673-688, 1992.

57. Apkon, M. and Nerbonne, J.M. α1-Adrenergic agonists selectively suppress voltage-dependent K+ currents in rat ventricular myocytes. Proc. Natl. Acad. Sci. USA 85:8756-8760, 1988.

58. Fedida, D., Braun, A.P. and Giles, W.R. α1-Adrenoceptors reduce background K+ current in rabbit ventricular myocytes. J. Physiol. 441:673-684, 1991.

59. Kushida, H., Hiramoto, T. and Endoh, M. The preferential inhibition of α1-over -adrenoceptor-mediated positive inotropic effect by organic calcium antagonists in the rabbit papillary muscle. Naunyn-Schmiedeberg's Arch. Pharmacol. 341: 206-214, 1990.

60. Gambassi, G., Spurgeon, H.A., Lakatta, E.G., Blank, P.S. and Capogrossi, M.C. Different effects of α- and -adrenergic stimulation on cytosolic pH and myofilament responsiveness to Ca^{++} in cardiac myocytes. Circ. Res. 71: 870-882, 1992.

61. Terzig A, Vogel SM. On the mechanism of the positive inotropic action of the alpha adrenoceptor agonist, phenylephrine, in isolated rat left atria. J Pharmacol Exp Ther 257:520-529, 1991.

62. Noland, T.A., Jr., Raynor, R.L. and Kuo, J.F. Identification of sites phosphorylated in bovine cardiac troponin I and troponin T by protein kinase C and comparative substrate activity of synthetic peptides containing the phosphory lation sites. J. Biol. Chem. 264:20778-20785, 1989.

63. Edes, I. and Kranias, E.G. Phospholamban and troponin I are substrates for protein kinase C *in vitro* but not in intact beating guinea pig hearts. Circ. Res. 67:394-400, 1990.

64. Yuan, S., Sunahara, F.A., Sen, A.K. Tumor-promoting phorbol esters inhibit cardiac functions and induce redistribution of protein kinase C in perfused beating rat heart. Circ. Res. 61:372-378, 1987.

65. Desemeci, A., Dhallan, R.S., Cohen, N.M., Lederer, W.J. and Rogers, T.B. Phorbol ester increases calcium current and simulates the effects of angiotensin II on cultured neonatal rat heart myocytes. Circ. Res. 62:347-357, 1988.

66. Capogrossi, M.C., Kaku, T., Filburn, C.R., Pelto, D.J., Hansford, R.G., Spurgeon, H.A. Lakatta, E.G. Phorbol ester and dioctanoylglycerol stimulate membrane association of protein kinase C and have a negative inotropic effect mediated by changes in cytosolic Ca^{++} in adult rat cardiac myocytes. Circ. Res. 66:1143-1155, 1990.

67. Gwathmay, J.K. and Hajjar, R.J. Effect of protein kinase C activation on sarcoplasmic reticulum function and apparent myofibrillar Ca^{++} sensitivity in intact and skinned muscles from normal and diseased human myocardium. Circ. Res. 67:744-752, 1990.

68. Teutsch, I., Weible, A. and Siess, M. Differential inotropic and chronotropic effects of various protein kinase C activators on isolated guinea pig atria. Eur. J. Pharmacol. 144:363-367, 1987.

69. MacLeod, K.T. and Harding, S.E. Effects of phorbol ester on contraction, intracellular pH and intracellular Ca^{++} in isolated mammalian ventricular myocytes. J. Physiol. 444:481-498, 1991.

70. Endoh, M., Otomo, J., Norota, I. and Takanashi, M. Selective inhibition by phorbol 12,13-dibutyrate of the α1-receptor-mediated positive inotropic effect. Internat. J. Cardiol. 40:191-201, 1993.

71. Endoh, M. and Takanashi, M. Differential inhibitory action of phorbol-12,13-dibutyrate on the positive inotropic effect of endothelin-1 and Bay K 8644 in the isolated rabbit papillary muscle. J. Cardiovasc. Pharmacol. 17(Suppl. 7): S165-S168, 1991.

72. Kushida, H., Hiramoto, T., Satoh, H. and Endoh, M. Phorbol ester does not mimic, but antagonizes the alpha-adrenoceptor-mediated positive inotropic effect in the rabbit papillary muscle. Naunyn-Schmiedeberg's Arch. Pharmacol. 337:169-176, 1988.

73. Endoh, M., Norota, I., Takanashi, M., Kasai, H. Inotropic effects of staurosporine, NA 0345 and H-7, protein kinase C inhibitors, on rabbit ventricular myocardium: selective inhibition of the positive inotropic effect mediated by α1-adrenoceptors. Jpn. J. Pharmacol. 63:17-26, 1993.

74. Zaza, A., Kline, R.P. and Rosen, M.R. Effects of α-adrenergic stimulation on intracellular sodium activity and automaticity in canine Purkinje fibers. Circ.Res. 66:416-426, 1990.

75. Tohse, N., Kameyama, M. and Irisawa, H. Intracellular Ca^{++} and protein kinase C modulate K+ current in guinea pig heart cells. Am. J. Physiol. 253:H1321-H 1324, 1987.

76. Kurachi, Y., Ito, H., Sugimoto, T., Miki, I. and Ui, M. α-Adrenergic activation of the muscarinic K+ channel is mediated by arachidonic acid metabolites. Pfl|gers Arch. 414:102-104, 1989.

77. Endoh, M., Hiramoto, T., Ishihata, A., Takanashi, M. and Inui, J. Myocardial α1-adrenoceptors mediate positive inotropic effect and changes in phosphatidylinositol metabolism: species differences in receptor distribution and the intracellular coupling process in mammalian ventricular myocardium. Circ. Res. 68:1179-1190, 1991a.

78. Yanagisawa, M., Kurihara, H., Kimura, S., Tomobe, Y., Kobayashi, M.,Mitsui Y., Yazaki, Y., Goto, K. and Masaki, T. A novel potent vasoconstrictor peptide produced by vascular endothelial cells. Nature 332:11-415, 1988.

79. Gu, X.H., Ou, R.C., Casley, D.J., Daly, M.J. and Nayler, W.G. Effect of aging on endothelin-1 binding sites in rat cardiac ventricular membranes. J. Cardiovasc. Pharmacol. 19:764-769, 1992.

80. Takanashi, M. and Endoh, M. Characterization of the positive inotropic effect of endothelin on mammalian ventricular myocardium. Am. J. Physiol. 261:H611-H619, 1991.

81. Moravec, C.S., Reynolds, E.E., Stewart, R.W. and Bond, M. Endothelin is a positive inotropic agent in human and rat heart *in vitro* Biochem. Biophys. Res. Commun. 159:14-18, 1989.

82. Ishikawa, T., Yanagisawa, M., Kimura, S., Goto, K. and Masaki, T. Positive inotropic action of novel vasoconstrictor peptide endothelin on guinea pig atria. Am. J. Physiol. 255:H970-H973, 1988.

83. Endoh, M., Shimizu, T. and Yanagisawa, T. Characterization of adrenoceptors mediating positive inotropic responses in the ventricular myocardium of the dog. Br. J. Pharmacol. 64:53-61, 1978.

84. Vigne, P., Lazdunski, M. and Frelin, C. The inotropic effect of endothelin-1 on rat atria involves hydrolysis of phosphatidylinositol. FEBS Let. 249:143-146, 1989.

85. Takanashi, M. and Endoh, M. Concentration- and time-dependence of phosphoinositide hydrolysis induced by endothelin-1 in relation to the positive inotropic effect in the rabbit ventricular myocardium. J. Pharmacol.Exp. Ther. 262:1189-1194, 1992.

86. Kramer, B.K., Smith, T.W. and Keely, R.A. Endothelin and increased contractility in adult rat ventricular myocytes: role of intracellular alkalosis induced by activation of the protein kinase C-dependent N+-H+ exchanger. Circ. Res. 68: 269-279, 1991.

87. Habuchi, Y., Tanaka, H., Furukawa, T., Tsujimura, Y., Takahashi, H. and Yoshimura, M. Endothelin enhances delayed potassium current via phospholipase C in guinea pig ventricular myocytes. Am. J. Physiol. 262:H345-H354, 1992.

88. Ishii, K., Numoki, K., Murakoshi, H. and Taira, N. Cloning and modulation by endothelin-1 of rat cardiac K channel. Biochem. Biophys. Res. Commun. 184:1484-1489, 1992.

89. Kim, D. Endothelin activation of an inwardly rectifying K+ current in atrial cells. Circ. Res. 69:250-255, 1991.

90. Tohse, N., Nakaya, H., Hattori, Y., Endou, M. and Kanno, M. Inhibitory effect mediated by $\alpha 1$-adrenoceptors on transient outward current in isolated rat ventricular cells. Pfl|gers Arch. 415:575-581, 1990.

91. Lauer, M.R., Gunn, M.D. and Clusin, W.T. Endothelin activates voltage-depen dent Ca^{++} current by a G protein-dependent mechanism in rabbit cardiac myocytes. J. Physiol. 448:729-747, 1992.

92. Wang, J.X., Paik, G. and Morgan, J.P. Endothelin-1 enhances myofilament Ca^{++}-respopnsiveness in aequorin-loaded ferret myocardium. Circ. Res. 69:582-589, 1991.

93. Masaki, T., Kimura, S., Yanagisawa, M. and Goto, K. Molecular and cellular mechanism of endothelin regulation: implication for vascular function. Circulation 84:1458-1468, 1991.

94. Kasai, H., Takanashi, M., Takasaki, C. and Endoh, M. Pharmacological characteristics of endothelin receptor subtypes mediating positive inotropic effects in rabbit heart. Am. J. Physiol., 1994 (in press)

95. Timmermans, P.B.M.W.M., Wong, P.C., Chiu, A.T., Herblin, W.F., Benfield, P., Carini, D.J., Lee, R.J., Wexler, R.R., Saye, J.A.M. and Smith, R.D. Angiotensin II receptors and angiotensin II receptor antagonists Pharmacol. Rev. 45:205-251, 1993.

96. Freer, R.J., Pappano, A.J., Peach, M.J., Bing, K.T., McLean, M.J., Vogel, S. and Sperelakis, N. Mechanism for the positive inotropic effect of angiotensin II on isolated cardiac muscle. Circ. Res. 39:178-182, 1976.

97. Baker, K.M. and Aceto, J.A. Characterization of avian angiotensin II cardiac receptors: coupling to mechanical activity and phosphoinositide metabolism. J. Mol. Cell. Cardiol. 21:375-382, 1989.

98. Kass, R.S. and Blair, M.L. Effects of angiotensin II on membrane current in cardiac Purkinje fibers. J. Mol. Cell. Cardiol. 13:797-809, 1981.

99. Rogers, T.B., Gaa, S.T. and Allen, I.S. Identification and characterization of functional angiotensin II receptors on cultured heart myocytes. J. Pharmacol. Exp. Ther. 236:438-444, 1986.

100. Kem, D.C., Johnson, E.I.M., Capponi, A.M., Chardonnens, D., Lang, U., Blondel, B., Koshida, H. and Vallotton, M.B. Effect of angiotensin II on cytosolic free calcium in neonatal rat cardiomyocytes. Am. J. Physiol. 261: C77-C85, 1991.

101. Moravec, C.S., Schluchter, M.D., Paranandi, L., Czerska, B., Stewart, R.W., Rosenkranz, E. and Bond, M. Inotropic effects of angiotensin II on human cardiac muscle *in vitro*. Circulation 82:1973-1984, 1990.

102. Meulemans, A.L., Andries, L.J. and Brutsaert, D.L. Does endocardial endothelium mediate positive inotropic response to angiotensin I and angiotensin II? Circ. Res. 66:1591-1601, 1990.

103. Kobayashi, M., Furukawa, Y. and Chiba, S. Positive chronotropic and inotropic effects of angiotensin II in the dog heart. Eur. J. Pharmacol. 50:17-25, 1978.

104. Zhang, J., Pfaffendorf, M. and van Zwieten, P.A. Positive inotropic action of angiotensin II in the pithed rat. Naunyn-Schmiederberg's Arch. Pharmacol. 347:658-663, 1993.

105. Baker, K.M. and Singer, H.A. Identification and characterization of guinea pig angiotensin II ventricular and atrial receptors: coupling to inositol phosphate production. Circ. Res. 62:896-904, 1988.

106. Feolde, E., Vigne, P. and Frelin, C. Angiotensin AT1 receptors mediate a positive inotropic effect of angiotensin II in guinea pig atria. Eur. J. Pharmacol. 245:63-66, 1993.

107. Wright, G.B., Alexander, R.W., Ekstein, L.S. and Gimborne, M.A., Jr. Characterization of the rabbit ventricular myocardial receptor for angiotensin II. Mol. Pharmacol. 24:213-221, 1983.

108. Baker, K.M., Campanile, C.P., Trachte, G.J. and Peach, M.J. Identification and characterization of the rabbit angiotensin II myocardial receptor. Circ. Res. 54:286-293, 1984.

109. Ishihata, A. and Endoh, M. Pharmacological characteristics of the positive inotropic effect of angiotensin II in the rabbit ventricular myocardium. Br. J. Pharmacol. 108:999-1005, 1993.

110. Endoh, M., Morita, H. and Kimura, J. Activation of chloride channel via AT1 angiotensin receptors in rabbit ventricular myocytes. Circulation 88 (suppl. 2): I-31, 1993.

111. Hofmann, P.A. and Fuchs, F. Effect of length and cross-bridge attachment on Ca^{++} binding to cardiac troponin C. Am. J. Physiol. 253:C90-C96, 1987.

112. Allen, D.G. and Kurihara, S. The effects of muscle length on intracellular calcium transients in mammalian cardiac muscle. J. Physiol. 327:79-94, 1982.

113. Housmans, P.R., Lee, N.K.M. and Blinks, J.R. Active shortening retards the decline of the intracellular calcium transient in mammalian heart muscle. Science 221:159-161, 1983.

114. Kawabata, Y. and Endoh, M. Effects of the positive inotropic agent Org 30029 on developed force and aequorin light transients in intactcanine ventricular myocardium. Circ. Res. 72:597-606, 1993.

115. Kitada, Y., Narimatsu, A., Matsumura, N. and Endo, M. Increase in Ca^{++} sensitivity of the contractile system by MCI-154, a novel cardiotonic agent, in chemically skinned fibers from the guinea pig papillary muscles J. Pharmacol. Exp. Ther. 243:633-638, 1987.

116. Endoh, M., Shibasaki, T., Satoh, H., Norota, I. and Ishihata, A. Different mechanisms involved in the positive inotropic effects of benzimidazole derivative UD-CG 115 BS (pimobendan) and its demethylated metabolite UD-CG 212 Cl in canine ventricular myocardium. J. Cardiovasc. Pharmacol. 17:365-375, 1991.

117. Solaro, R.J. and R|egg, J.C. Stimulation of Ca^{++} binding and ATPase activity of dog cardiac myofibrils by AR-L 115 BS, a novel cardiotonic agent. Circ. Res. 51:290-294, 1982.

118. Endoh, M., Yanagisawa, T., Morita, T. and Taira, N. Differential effects of sulmazole (AR-L 115 BS) on contractile force and cyclic AMP level in canine ventricular muscle: comparison with MDL 17,043. J. Pharmacol. Exp. Ther. 234:267-273, 1985.

119. Fabiato, A. and Fabiato, F. Techniques of skinned cardiac cells and of isolated cardiac fibers with disrupted sarcolemmas with reference to the effects of catecholamines and of caffeine. In: Recent Adv. Stud. Card. Struct. Metab., eds. Roy, P.E. and Dhalla, N.S. vol 9:71-94, 1976, University Park Press, Baltimore.

120. Wendt, I.R. and Stephenson, D.G. Effects of caffeine on Ca-activated force production in skinned cardiac and skeletal muscle fibres of the rat. Pfl|gers Arch. 398:210-216, 1983.

121. Grob, T., Lues, I and Daut, J. A new cardiotonic drug reduces the energy cost of active tension in cardiac muscle. J. Mol. Cell. Cardiol. 25:239-244, 1993.

122. Lee, J.A. and Allen, D.G. EMD 53998 sensitizes the contractile proteins to calcium in intact ferret ventricular muscle. Circ. Res. 69:927-936, 1991.

123. Gambassi, G., Capogrossi, M.C., Klockow, M. and Lakatta, E.G. Enant iomeric dissection of the effects of the inotropic agent, EMD 53998, in single cardiac myocytes. Am. J. Physiol. 264:H728-H738, 1993.

124. White, J., Lee, J.A., Shah, N. and Orchard, C.H. Differential effects of the optical isomers of EMD 53998 on contraction and cytoplasmic Ca^{++} in isolated ferret cardiac muscle. Circ. Res. 73:61-70, 1993.

125. Endoh, M. The effects of theophylline on aequorin light transients and force in the isolated dog right ventricular myocardium. J. Mol. Cell. Cardiol. 26:87-98, 1994.

126. Abe, Y., Kitada, Y. and Narimatsu, A. Beneficial effect of MCI-154, a cardiotonic agent, on ischemic contractile failure and myocardial acidosis of dog hearts: comparison with dobutamine, milrinone and pimobendan. J. Pharmacol. Exp. Ther. 261:1087-1095, 1992.

127. Strauss, J.D., Zeugner, C. and R|egg, J.C. The positive inotropic calcium sensitizer EMD 53998 antagonizes phosphate action on cross-bridges in cardiac skinned fibers. Eur. J. Pharmacol. 227:437-441, 1992.

15. Effect of Inotropic Agents on Mechanoenergetics in Human Diseased Heart

Motoshi Takeuchi, Hideyuki Takaoka, Katsuya Hata, Masuki Mori, Hideyuki Yamakawa, Mitsuhiro Yokoyama

INTRODUCTION

For the treatment of heart failure, inotropic agents yield short-term hemodynamic improvement. However, inotropic therapy is still an unfilled promise for the treatment of chronic heart failure (1). This may partly result from the fact that inotropic interventions increase myocardial oxygen consumption (VO_2). Therefore, the investigation of myocardial energetics in relation to inotropic agents in patients with heart disease has clinical and therapeutic relevance. Recently, the concept of left ventricular (LV) elastance (E_{max}) and systolic pressure-volume area (PVA) has been proposed by Suga and his colleagues (2,3,4,5). Experimental studies have shown that the relation between VO_2 and PVA is linear under a variety of loading conditions. This linear relation has been used to partition VO_2 into non-mechanical and mechanical portions. Experimental studies have shown that positive inotropic interventions, which enhance E_{max}, shift the VO_2-PVA relation line upward in a parallel manner. The resultant increase in VO_2 intercept of the relation has been considered to be VO_2 for non-mechanical work. The relation between E_{max} and PVA-independent VO_2 has been reported to be linear; the slope represents "oxygen cost of contractility" under various inotropic interventions. The aim of this study was to examine the effects of various inotropic agents on human cardiac mechanoenergetics by utilizing the concept of E_{max} and the VO_2-PVA relation. To this end, we compared the effects of the newly developed phosphodiesterase inhibitor E-1020 and a novel Ca^{++} sensitizing agent, MCI-154, with dobutamine in patients with heart disease.

Measurement of LV Volume

Measurement of LV volume is essential for understanding and assessing LV function in humans. A number of methods to measure LV volume using cine-angiography, radionuclide ventriculography, echocardiography and nuclear magnetic resonance have been developed. However, most of them have significant limitations. Recently, a new method that combines the measurement of LV pressure and volume by means of a conductance catheter has been introduced by Baan and colleagues (6). A previous experimental study has shown the accuracy of conductance-derived LV volume in comparison with LV volume assessed by the intracavity balloon method in isolated, ejecting dog hearts (7). This method employs a multi-electrode catheter to measure intracavity electrical conductance from which LV volume is calculated. To generate an electrical field (2kHz,30μA) in the LV, the conductance catheter with 10 electrodes is connected to a stimulator/processor (Sigma-5, Cardiodynamics, Rijnsberg, Netherlands). The six interposed electrodes are used to measure the conductances of five LV segments. The electrode distance was chosen according to the LV volume, such that the most proximal electrode is situated just above the aortic valve with the distal electrode in the apex. Instantaneous LV volume, V(t), is computed as

$$V(t) = 1/\alpha \cdot [L2 \cdot \rho \cdot G(t) - \alpha Vc]$$

where G(t) is the instantaneous sum of the conductances, L is the distance between two adjacent electrodes, ρ is blood resistivity in the LV, $1/\alpha$ is the gain constant and αVc is the correcting volume for the conductance of the surrounding tissues. Blood

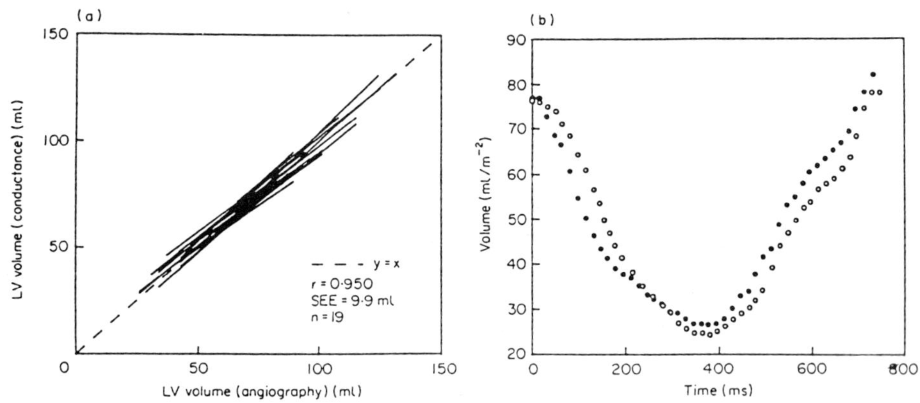

Fig. 1a. Comparison of left ventricular volume during cardiac cycle in each patient between conductance catheter and angiography. b. Illustration of volume-time plots constructed from the conductance volume and cineangiography in a representative case.

resistivity of the LV was measured in a calibrated cuvette. $1/\alpha$ was determined individually by obtaining the ratio of stroke volume determined by the thermodilution method to that measured by the conductance catheter. αVc was estimated by bolus injection of 10 mL of hypertonic saline (5% NaCl) into the pulmonary artery.

Validation of Volume Measurement by the Conductance Catheter Method

To validate the accuracy of human LV volume measured by the conductance catheter method, conductance volume was compared with LV volume measured by biplane angiography in patients (8). The regression lines between conductance and cine-angiographic volumes throughout one cardiac cycle are shown in Figure 1a. The slopes were close to unity and intercepts were very small. The correlation coefficients were very high and there were no significant differences in the correlation coefficients between normal and depressed hearts. Figure 1b shows representative volume-time plots constructed from conductance and cine-angiography. The shape of volume-time plots by the conductance method was quite similar to those by cine-angiography. Thus, the conductance catheter method can accurately and continuously estimate the volume cycle in human heart.

ASSESSMENT OF VENTRICULAR FUNCTION IN HUMAN DISEASED HEART

LV Contractility

Cardiac catheterization was performed on patients in the fasting state via the femoral approach. An 8F conductance catheter (CardioDynamics, Rijnsberg, Netherlands) and a 2F micro-tip catheter pressure transducer (Millar Instruments Inc., Tx) were inserted into the LV to obtain instantaneous LV pressure and volume data. A 7F Swan-Ganz thermodilution catheter (Goodtech, Inc., PA) was advanced to the pulmonary artery for the measurement of cardiac output and the injection of hypertonic saline.

A large balloon occlusion catheter (Baxter, Inc., CA) was advanced to the right atrium to perform inferior vena caval occlusion. LV pressure-volume (P-V) relations were simultaneously determined by the conductance catheter. ESPVR was calculated from variably loaded P-V loops during the inferior vena caval occlusion (9,10).

Figure 2a shows continuous on-line monitoring of the volume and pressure during inferior vena caval occlusion by the conductance catheter method. Figure 2b shows the P-V diagrams and ESPVR.

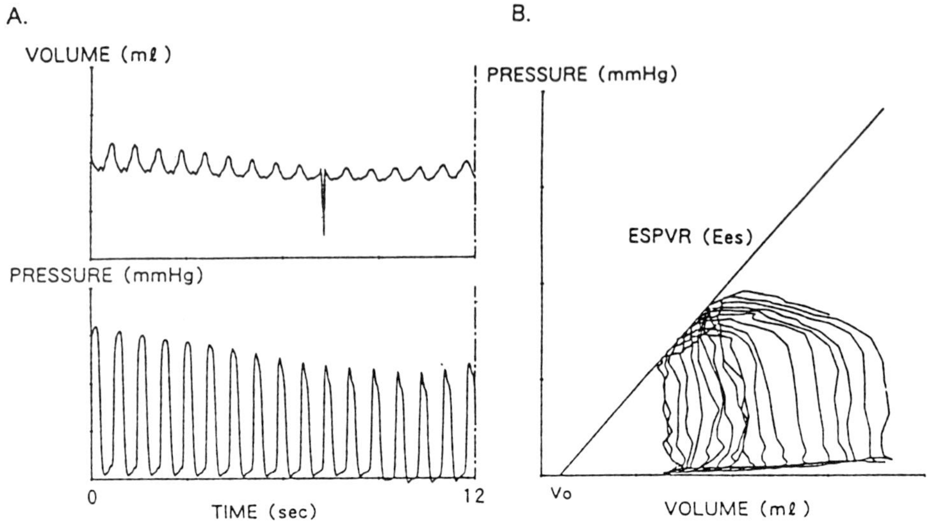

Fig. 2a. Continuous on-line recordings of left ventricular volume and pressure during inferior vena caval occlusion. 2b. Left ventricular pressure-volume diagram and end-systolic pressure-volume relation.

Arterial-Ventricular Coupling

Following the approach proposed by Sunagawa et al (11), we determined the effective arterial elastance (E_a), a parameter that incorporates the values of Windkessel model elements and heart rate, as the ratio of the end-systolic pressure to stroke volume. E_a is the negative value of the slope of the diagonal line connecting the end-systolic pressure-volume point and the end-diastolic point on the volume axis. E_a/E_{max} represents arterial-ventricular coupling. Analysis of arterial-ventricular coupling is useful to evaluate the relation between LV contractility and arterial afterload. Coupling analysis may have the potential to discriminate between the positive inotropic and vasodilative effects of inotropic agents on cardiac output. Furthermore, experimental studies have demonstrated the relation between arterial-ventricular coupling and mechanical efficiency, defined as the ratio of external work (EW) to VO_2 (12). Figure 3 shows a representative P-V diagram and the method of calculating E_a and EW. EW was defined as the area in the P-V loop for each cardiac cycle.

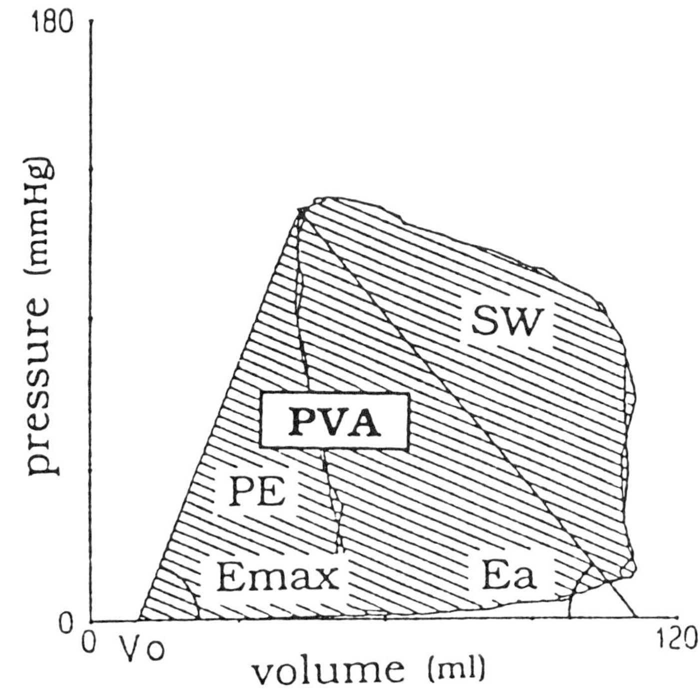

Fig. 3. Representative pressure-volume diagram showing the method of calculating systolic pressure-volume area (PVA).

Measurement of Myocardial Oxygen Consumption and Mechanical Efficiency

A Webster catheter (Wilton Webster Manufacturing Co., CA) was advanced into the coronary sinus via the left subclavian artery and coronary sinus blood flow (CSF) was measured during a 30-second continuous injection of 5% glucose at room-temperature through the catheter lumen at a rate of 40 mL/min with use of a Mark IV angiographic injector (Medral Inc., PA). Coronary venous blood was sampled from the distal lumen of the Webster catheter. VO_2 per minute was calculated as the product of CSF (mL/min) and coronary arteriovenous oxygen content difference (vol %), divided by heart rate (9,10). LV mechanical efficiency was calculated as the ratio of EW (in joules per beat) to VO_2 (in joules per beat), where 1 mmHg·mL of EW and 1 mL O_2 of VO_2 correspond to 0.000133 and 20 joules, respectively (10,13,14).

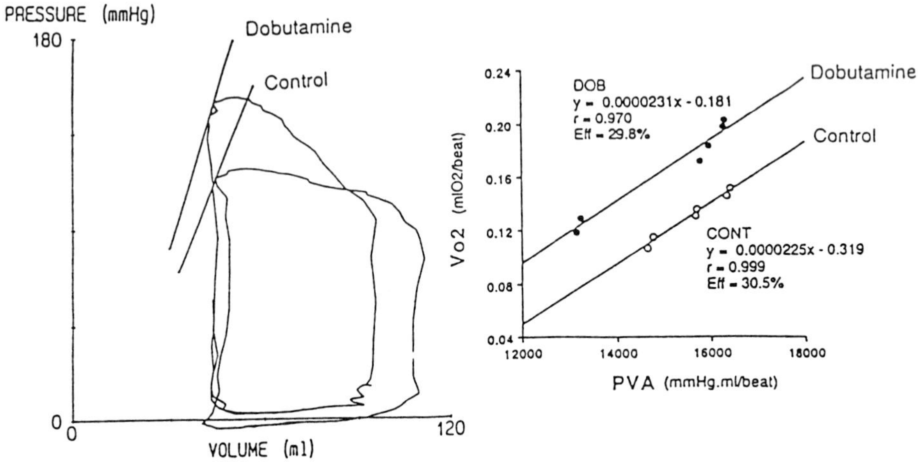

Fig. 4. Representative pressure-volume diagram and VO_2-PVA relations in control and enhanced contractile state.

VO_2-PVA Relation and Oxygen Cost of Contractility

Following the approach of Suga and his colleagues (2,3,4,5), PVA was defined as the area enclosed by the ESPVR line, the ESPVR curve and the systolic trajectory of the P-V loop (9,10).

We assessed the relation between VO_2 and PVA in patients with heart disease during volume loading. The VO_2 and PVA data during volume loading were fit to a linear regression line. PVA correlated linearly with VO_2 under control conditions (9) (see Figure 4). After dobutamine infusion, PVA-independent VO_2 increased and the VO_2-PVA relation was elevated in a parallel manner. Thus, PVA is a reliable and valuable predictor of VO_2 under different contractile states. The increase in the VO_2 intercept represents VO_2 for increased excitation-contraction coupling associated with

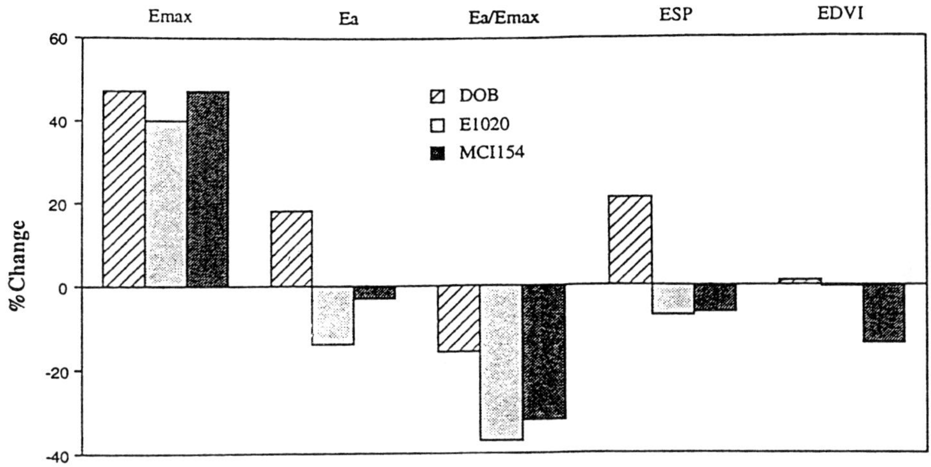

Fig. 5. Effect of dobutamine, phosphodiesterase inhibitor E1020 and Ca^{++} sensitizing inotropic agent MCI-154 on hemodynamics. E_{max} = slope of end-systolic pressure-volume relation, E_a = arterial effective elastance, E_a/E_{max} = arterial-ventricular coupling, ESP = end-systolic pressure. EDVI = end-diastolic volume index.

the augmented contractile state. The parallelism of the VO_2-PVA relation discriminates between VO_2 for PVA-dependent VO_2 (VO_2 for mechanical work) and VO_2 for PVA-independent VO_2. The PVA-independent VO_2 of a contraction at enhanced E_{max} was calculated as VO_2 minus PVA-dependent VO_2 of the contraction. The oxygen cost of contractility was calculated as the ratio of the increase in PVA-independent VO_2 to the increase in E_{max}.

Effect of Inotropic Agents on Mechanics and Myocardial Energetics in Human Diseased Heart

Vasodilator therapy has gained widespread acceptance for the treatment of heart failure. Reduction of increased systemic vascular resistance in congestive heart

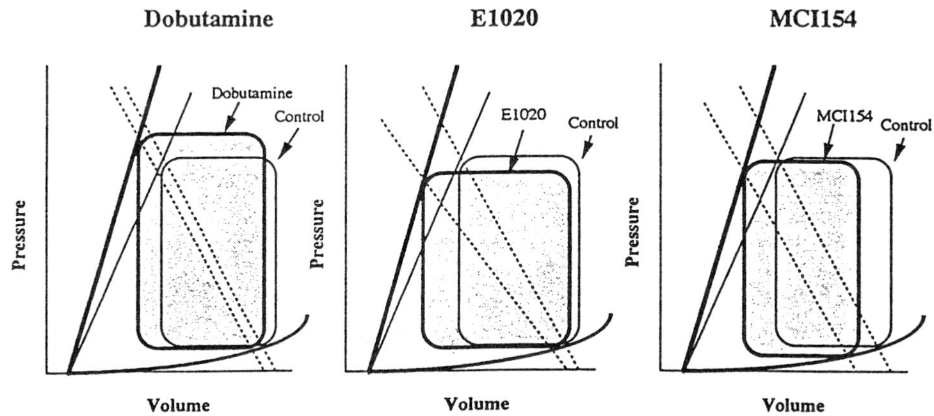

Fig. 6. Scheme of effect of dobutamine phosphodiesterase inhibitor E1020 and Ca^{++} sensitizing inotropic agent MCI-154 on pressure-volume parameters.

failure improves cardiac output and reduces myocardial oxygen consumption. Phosphodiesterase-inhibitors are known to have positive inotropic effects (15). On the other hand, a positive inotropic action induced by altering the responsiveness of the myofilaments to Ca^{++} would be expected to increase contractile state with a lower myocardial oxygen consumption compared to a positive inotropic action due to increasing $(Ca^{++})_i$ (16). Recently, a novel cardiotonic agent, MCI-154, has been reported to exert a positive inotropic action by altering the responsiveness of the myofilament to calcium (17). We compared the effects of an inotropic agent without Ca^{++} sensitizing effect, dobutamine, a PDE-inhibitor (E1020), and the Ca^{++} sensitizing agent MCI-154 on mechanics and energetics in the diseased human heart. Figures 5 and 6 illustrate the effects of inotropic agents on hemodynamic variables. E_{max} increased to the same extent with the three inotropic agents. Arterial afterload, E_a, was unchanged with dobutamine and MCI-154, but decreased with PDE-inhibitor. Thus, E_a/E_{max} decreased with all inotropic agents but the extent of reduction was greater with the PDE-inhibitor than with dobutamine and MCI-154.

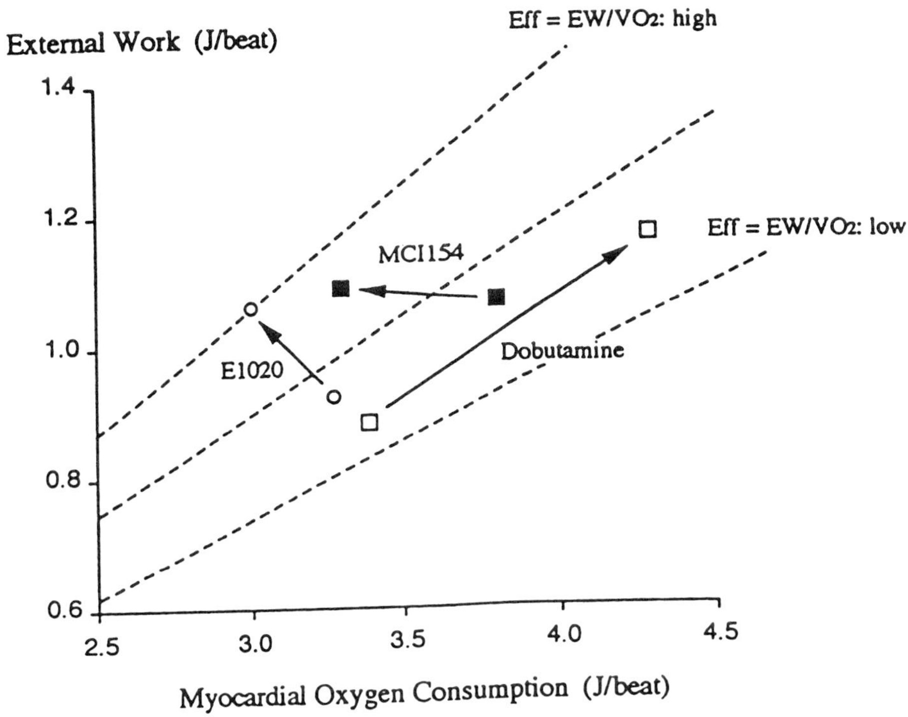

Fig. 7. Effects of dobutamine, phosphodiesterase inhibitor E1020 and Ca^{++} sensitizing inotropic agent MCI-154 on the relation between PVA-independent VO_2 and E_{max}. The oxygen wasting cost of contractility was calculated as the ratio of the increase in PVA-independent VO_2 to the increase in E_{max}.

Figure 7 illustrates the effects of the inotropic agents on the relation between EW and VO_2. Dobutamine increased VO_2 in proportion to the increase in EW. On the other hand, the PDE-inhibitor increased EW without an increase in VO_2. After administration of MCI-154, VO_2 decreased without an increase in EW. Therefore, mechanical efficiency did not improve with dobutamine, but improved with the PDE-inhibitor and MCI-154.

Figure 8 illustrates the relation between the PVA-independent VO_2 and E_{max} before and after inotropic agents. When the inotropic agents increased E_{max} comparably, the PVA-independent VO_2 increased to the same extent with dobutamine and the

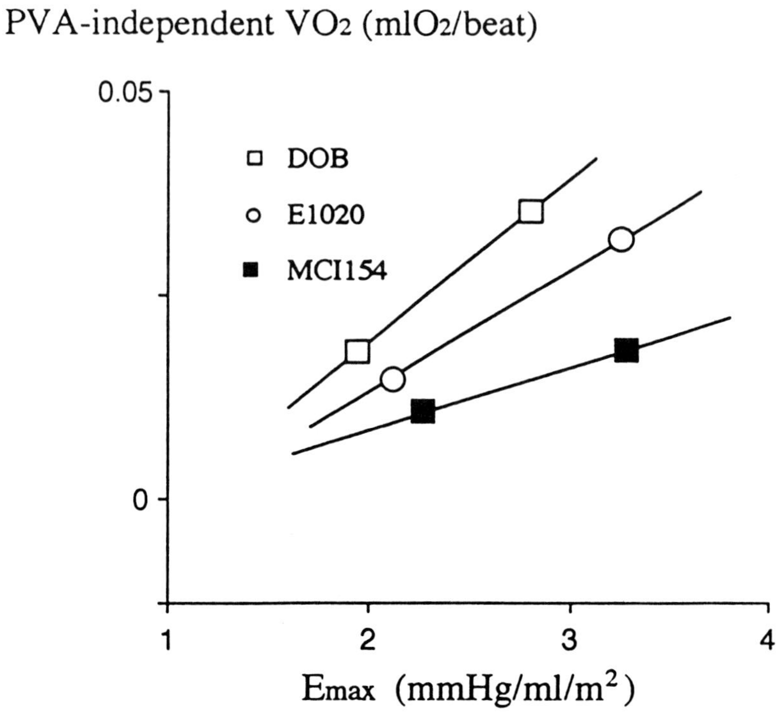

Fig. 8. Effects of dobutamine, phosphodiesterase inhibitor E1020 and Ca^{++} sensitizing inotropic agent MCI-154 on the relation between PVA-independent VO_2 and E_{max}. The oxygen wasting cost of contractility was calculated as the ratio of the increase in PVA-independent VO_2 to the increase in E_{max}.

PDE-inhibitor. On the other hand, the increase in PVA-independent VO_2 was smaller with MCI-154 than with the other agents. The oxygen wasting cost of contractility was calculated by the ratio of the increase in PVA-independent VO_2 to the increase in E_{max}. Therefore, the oxygen cost of contractility with dobutamine was comparable to the PDE-inhibitor, but that with MCI-154 was lower than with the other agents. Thus, the improvement of mechanical efficiency with the PDE-inhibitor may be responsible for the improvement of arterial-ventricular coupling. On the other hand, MCI-154 did not decrease E_a but improved arterial-ventricular coupling and mechanical efficiency. The improvement of mechanical efficiency with MCI-154 may not result from alterations in loading conditions, but from the lowered oxygen cost of contractility. Therefore, MCI-154 was superior to the other agents with regard to oxygen saving effects and these effects of a Ca^{++}-sensitizer may have beneficial effects on myocardial energetics in the diseased human heart. In conclusion, the concept of E_{max} and PVA can be applied to human heart and this analysis may be able to provide a better understanding of mechanical and energetic aspects of inotropic agents for the treatment of heart failure.

ACKNOWLEDGEMENTS

We gratefully appreciate the continuous encouragement by Prof. Hiroyuki Suga of the Second Department of Physiology, Okayama University Medical School, Okayama, Japan.

REFERENCES

1. Shah, P.K., et al. Inotropic therapy for refractory congestive heart failure with oral enoximone (MDL017,043): poor long-term results despite early hemodynamic and clinical improvement. Circulation 71:326-331, 1985.

2. Suga, H. Ventricular energetics. Physiol Rev 70:247-277, 1990.

3. Suga, H., et al. Effect of positive inotropic agents on the relation between oxygen consumption and systolic pressure volume area in canine left ventricle. Circ Res 53:306-318, 1983.

4. Goto, Y., et al. Effects of amrinone and isoproterenol on mechanoenergetics of blood-perfused rabbit heart. Am J Physiol 262:H719-H727, 1992.

5. Hata, K., et al. Mechanoenergetic effects of pimobendan in canine left ventricles: comparison with dobutamine. Circulation 86:1291-1301, 1992.

6. Baan, J., et al. Continuous measurement of left ventricular volume in animals and humans by conductance catheter. Circulation 70:812-823, 1984.

7. Burkhoff, D., et al. Accuracy of volume measurement by conductance catheter in isolated, ejecting canine hearts. Circulation 72:440-447, 1985.

8. Odake, M., et al. Determination of left ventricular volume using a conductance catheter in the diseased human heart. Eur Heart J 13(Suppl E):22-27, 1992.

9. Takaoka, H., et al. Comparison of hemodynamic determinants for myocardial oxygen consumption under different contractile states in human ventricle. Circulation 87:59-69, 1993.

10. Takaoka, H., et al. Comparison of the effects of arterial-ventricular coupling between phosphodiesterase inhibitor and dobutamine in the diseased human heart. J Am Coll Cardiol 22:598-606, 1993.

11. Sunagawa, K., et al. Optimal arterial resistance for the maximal stroke work studied in isolated canine left ventricle. Circ Res 56:586-595, 1985.

12. Burkhoff, D., Sagawa, K. Ventricular efficiency predicted by an analytical model. Am J Physiol 250:R1021-R1027, 1986.

13. Nichols, A.B., et al. Left ventricular mechanical efficiency in coronary artery disease. J Am Coll Cardiol 7:270-279, 1986.

14. Kameyama, T., et al. Energy conversion efficiency in human left ventricle. Circulation 85:988-996, 1992.

15. Mancini, D., et al. Intravenous use of amrinone for the treatment of the failing heart. Am J Cardiol 9:849-857, 1985.

16. Lee, J.A., Allen, D.G. Altering the strength of the heart: basic mechanisms in modulation of cardiac calcium sensitivity: A New Approach to Increasing the Strength of the Heart. Oxford University Press, New York 1-36, 1993.

17. Kitda, Y., et al. Increase in Ca^{++} sensitivity of the contractile system by MCI-154, a novel cardiotonic agent, in chemically skinned fibers from the guinea pig. J Pharmacol Exp Ther 243:633-638, 1987.

16. Effects of Various Inotropic Agents on the Relation Between Ventriculoarterial Coupling and Myocardial Energetics in Patients With Idiopathic Dilated Cardiomyopathy

Mitsuhiro Yokota, Hitoshi Ishihara, and Toshikazu Sobue

INTRODUCTION

Idiopathic dilated cardiomyopathy (DCM) is one of the fatal heart diseases which severely compromises cardiac performance. Various aspects of this disease have been analyzed and reported (1-7) including the pathophysiology, hemodynamics, and mortality, but many problems remain unresolved. One problem is that this disease is extremely resistant to therapy and no cardiotonic agent has been reported to decrease its mortality (8-11).

Recent studies on myocardial energetics using pressure-volume relationships have enhanced our understanding of the efficiency of energy conversion from oxygen consumption to mechanical or hydraulic output, and how the contractile state and loading conditions influence the efficiency (12-17). Mechanical efficiency, a ratio of external work (EW) to myocardial oxygen consumption (MVO_2), is the conversion efficiency from metabolic energy to the hydraulic energy that the left ventricle generates against the vascular system. This efficiency can be divided into two steps: 1) the efficiency from myocardial oxygen consumption to the pressure-volume area (PVA), and 2) the efficiency from pressure-volume area to external work which is called the cardiac work efficiency (12-14). Each step of energy transfer varies according to changes in contractility and the loading conditions. Previous studies have reported two criteria for optimal coupling between contractility and the loading conditions (18,19): 1) external work is maximized when effective arterial elastance (Ea) equals end-systolic elastance (Ees), and 2) the mechanical efficiency is maximized when Ea is nearly one half of Ees.

By means of left ventricular pressure-volume relationships, we investigated left ventricular contractility, the arterial loading conditions, and how their interaction affects myocardial energetics in patients with DCM.

METHODS

Patients

The study group consisted of 5 patients (4 men and 1 woman) with normal left ventricular ejection fraction for control subjects (mean age, 53 ± 13 years), and 31 patients (25 men and 6 women) with mild to moderate idiopathic dilated cardiomyopathy (mean age, 51 ± 13 years). All patients were in normal sinus rhythm. DCM was defined as a decreased left ventricular ejection fraction (<50%, as determined by contrast ventriculography) and a dilated left ventricular cavity in the absence of coronary or valvular heart disease, arterial hypertension, or cardiac muscle disease caused by any known systemic disease (20,21). All patients with DCM had experienced cardiac failure at least once and they complained of dyspnea at rest or with exertion and had pulmonary congestion and an increased cardiothoracic ratio. The study protocol was reviewed and approved by the institutional committee on human investigations, and each patient gave written informed consent before entering the study.

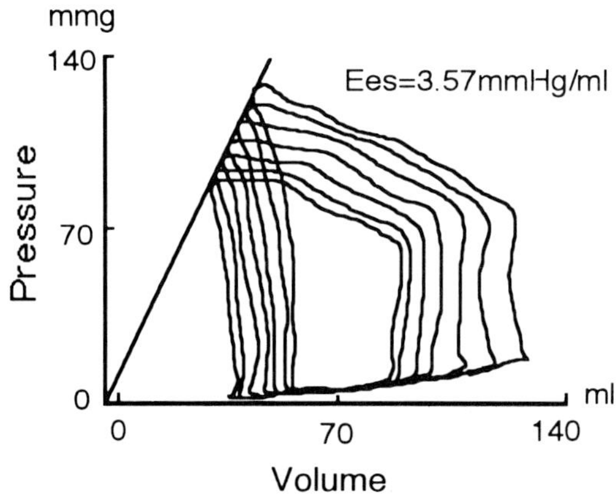

Fig. 1. End-systolic pressure-volume relation (Ees) was determined by transient preload reduction by occluding the inferior vena cava.

Study Protocol

All cardiac catheterizations were begun in the morning with the patients in a fasting state. All previous medications were withheld for at least 48 hours before the investigation. Catheterization of the right and left sides of the heart was performed by the femoral approach. A double-thermistor catheter (model CCS/GOK, Wilton-Webster Laboratories, Los Angeles, California) was placed percutaneously in the coronary sinus through the brachial or the subclavian vein to measure myocardial blood flow. After left ventriculography was performed, an 12-electrode volume conductance catheter (Leycom, Oegstgeest, Netherlands) with a micromanometer-tip catheter was advanced to the left ventricular apex to measure the left ventricular volume and pressure (22). A Fogarty catheter (model 62-080-8/22F, Edwards Laboratories, Santa Anna, California) was placed in the inferior vena cava and was inflated transiently with 8 to 15 ml of CO2 gas for the determination of the left ventricular end-systolic pressure-volume relationship (Figure 1). Baseline hemodynamics were obtained and myocardial blood flow and arterial and coronary sinus blood oxygen saturation were measured. A left ventricular pressure-volume diagram was obtained during the steady state and transient preload reduction by occluding the inferior vena cava. The ventricular volume obtained by conductance methods was calibrated by using bi-plane ventriculography [area length methods (23)] as follows. Bi-plane ventriculography was performed immediately before inserting the conductance catheter. A mesh grid placed at the same distance as the body from the image intensifier was also exposed bi-directionally for each case, and was used for the correction factor of ventricular volume. Both the end-systolic and end-diastolic volumes obtained by the conductance method were adjusted directly for each case by the end-systolic and end-diastolic volumes calculated from bi-plane ventriculography (24).

Myocardial blood flow was measured using a double-thermistor catheter which was positioned in the coronary sinus. Coronary sinus and great cardiac vein flows were calculated according to the method of Ganz et al. (25) using a computer system (Thermo Flow, Goodman Co., Ltd., Nagoya, Japan). Oxygen saturation was measured in samples obtained simultaneously from the femoral artery and the coronary sinus, while the steady state measurements were being collected.

Investigation of the Effects of Inotropic Agents

After the basal measurements were complete, three inotropic agents with different mechanisms of action were administered to patients with DCM. Basal characteristics of three groups receiving inotropic agents are provided in Table 1. There were no significant difference in sex, age, left ventricular end-diastolic volume index and left ventricular ejection fraction among these three groups. Dobutamine was administered intravenously (2.5 to 5.0 μg/kg/min) to 10 patients, 10 patients received 10 to 15 mg of the oral phosphodiesterase inhibitor MS-857 (26,27), and 7 patients received OPC-18790 intravenously (5.0 to 10 μg/kg/min). OPC-18790 is a drug having multiple mechanisms of action. It works mainly by inhibiting phosphodiesterase, and prolongs the action potential of myocytes. Hemodynamic measurements, coronary blood flow, and the pressure-volume diagram were obtained and blood sampling was performed: 1) when the hemodynamic indices stabilized 15 minutes after the continuous infusion of dobutamine; 2) 45 minutes after the oral administration of MS-857; or 3) 60 min after the continuous infusion of OPC-18790.

Table 1. Groups receiving inotropic agents.

Group	Dobutamine	MS-857	OPC-18790
N	10	10	7
Male/Female	8/2	8/2	5/2
Age	55 ± 8	52 ± 13	56 ± 13
LVEDV (ml)	237 ± 52	197 ± 58	215 ± 47
LVEF	0.30 ± 0.12	0.33 ± 0.09	0.38 ± 0.12
Dosage	2.5 to 5 (μg/kg/min)	10 to 15 (mg,oral)	5 to 10 (μg/kg/min)

Mean $\pm$ SD

VENTRICULOARTERIAL COUPLING AND MYOCARDIAL ENERGETICS IN PATIENTS WITH DCM

Baseline data from 5 control subjects and 31 patients with DCM are provided in Table 2. Left ventricular end diastolic volume index was larger and left ventricular ejection fraction was smaller in patients with DCM than in control subjects. Peak positive dP/dt and end systolic elastance were remarkably smaller in patients with DCM. Peak negative dP/dt was smaller and the time constant of left ventricular

Table 2. Clinical characteristics and hemodynamic data.

	Control (n=5)	DCM (n=31)	p Value
Age(years old)	53 ± 13	51 ± 13	NS
Sex(Male/Female)	4/1	25/6	NS
Heart rate(beats/min)	68 ± 11	73 ± 11	NS
Cardiac output(L/min)	4.8 ± 0.9	4.2 ± 0.9	NS
LVEDVI (ml/m^2)	66 ± 14	132 ± 40	***
LV ejection fraction(%)	71 ± 6	36 ± 12	****
LVEDP (mmHg)	13 ± 3	13 ± 6	NS
Peak(+)dp/dt(mmHg/sec)	2049 ± 151	1092 ± 272	****
Peak(-)dp/dt(mmHg/sec)	-2226 ± 298	-1168 ± 286	****
Tau (Weiss,msec)	38 ± 15	51 ± 12	**
Ees (mmHg/ml)	2.9 ± 1.3	0.9 ± 0.5	****
Ea/Ees	1.0 ± 0.2	3.2 ± 0.3	****

LVEDVI:left ventricular end-diastolic volume index,LVEDP:left ventricular end-diastolic pressure,Tau:time constant of isovolumic pressure decay,Ees:end-systolic elastance,Ea:effective arterial elastance, $^{**}p<0.01$, $^{***}p<0.001$, $^{****}p<0.0001$, Mean value $\pm$ SD isovolumic pressure decay (Weiss et al.) was larger in patients with DCM. These results indicated that both systolic and diastolic function were severely impaired in patients with DCM.

Ea/Ees was almost 1.0 in control subjects. This meant the hearts with normal cardiac function had almost the optimal condition with matched contractile state and arterial loading condition. On the other hand, patients with DCM had a large Ea and a small Ees, and the coupling condition was far from the optimal condition.

Figure 2 shows representative sets of left ventricular pressure-volume diagrams obtained from a control subject and a DCM patient. The difference in the ventriculoarterial coupling can be easily seen in these diagrams.

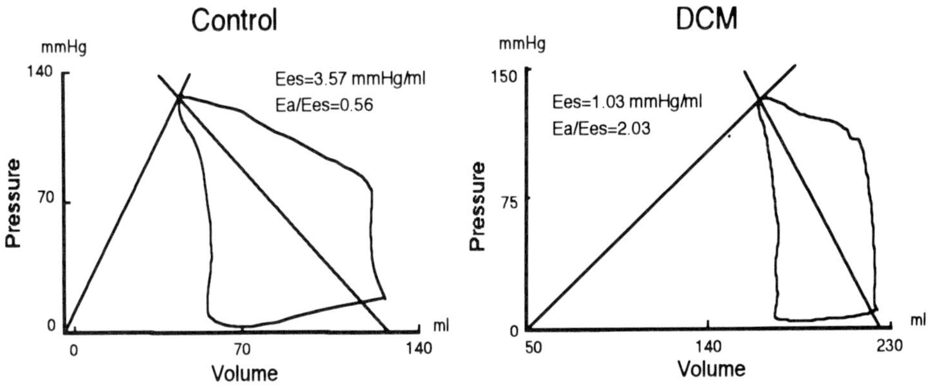

Fig. 2. Left ventricular pressure-volume relation in a control subject and a patient with DCM. Ventriculoarterial coupling was almost optimal in a control subject, while it was far from optimal in a patient with DCM.

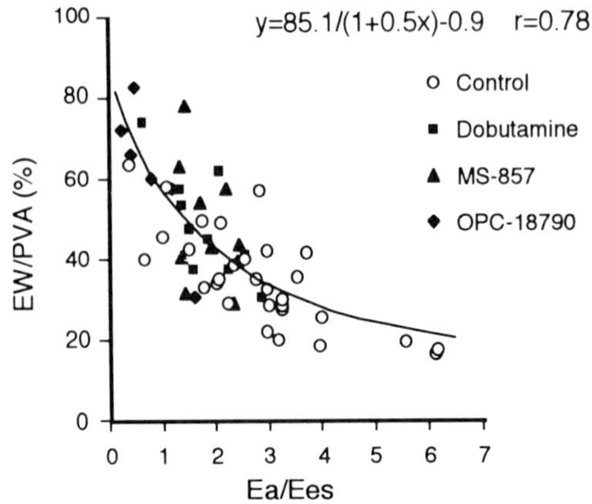

Fig. 3. Work efficiency had a hyperbolic correlation to Ea/Ees. The obtained equation resembled the equation derived theoretically (EW/PVA=1+0.5Ea/Ees).

Figure 3 shows the scatter diagram of cardiac work efficiency against the ratio of Ea to Ees for all the data obtained at baseline and after the administration of all drugs. A hyperbolic correlation between cardiac work efficiency and the ratio of Ea to Ees was found with a correlation coefficient of 0.78 ($p<0.001$). The experimentally obtained regression equation closely resembled the equation which was derived theoretically (EW/PVA=1+0.5Ea/Ees).

Figure 4 shows the close positive correlation between the ratio of PVA to MVO_2 and the ratio of Ea to Ees in all the data. The correlation coefficiency was 0.66 ($p<0.001$).

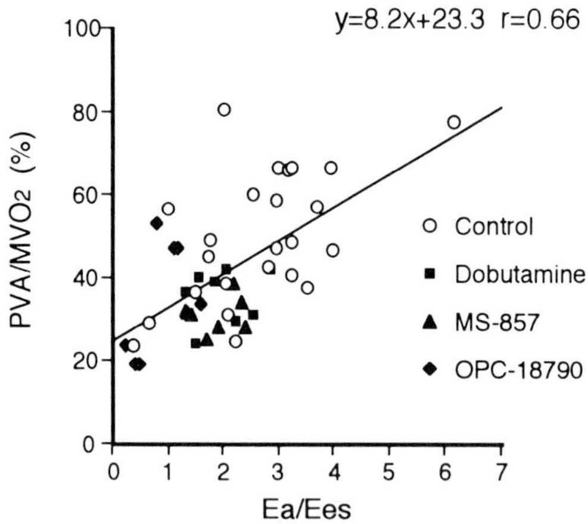

Fig. 4. The first stage of energy transfer from metabolic energy to mechanical energy (PVA/MVO_2) had a positive correlation to Ea/Ees.

As shown in Figure 5, the mechanical efficiency remained within a narrow range depending on the interaction of the first and second stage of energy transfer.

EFFECTS OF INOTROPIC AGENTS ON HEMODYNAMICS AND MYOCARDIAL ENERGETICS IN PATIENTS WITH DCM

Effects on Ventriculoarterial Coupling and Myocardial Energetics

For the group receiving the dobutamine infusion, the baseline Ees increased from 0.80 ± 0.11 to 1.31 ± 0.13 mm Hg/ml ($p<0.01$), Ea increased slightly from 2.14 ± 0.19 to 2.33 ± 0.22 mm Hg/ml ($p<0.01$), and Ea/Ees decreased from 3.12 ± 0.43 to 1.86 ± 0.15 ($p<0.05$). For the group receiving MS-857, the baseline Ees increased from 1.03 ± 0.14 to 1.39 ± 0.17 mm Hg/ml ($p<0.01$), Ea decreased from 2.87 ± 0.36 to 2.32 ± 0.31 mm Hg/ml ($p<0.05$), and Ea/Ees decreased from 3.16 ± 0.45 to 1.78 ± 0.22 ($p<0.01$). For the group receiving OPC-18790, the baseline Ees increased from 1.80 ± 0.38 to 2.68 ± 0.57 mm Hg/ml ($p<0.01$), Ea did not change significantly, and Ea/Ees decreased from 1.49 ± 0.31 to 0.84 ± 0.18 ($p<0.05$).

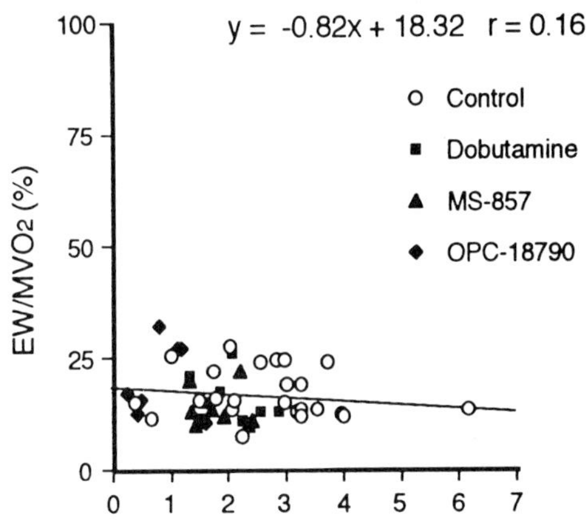

Fig. 5. Mechanical efficiency remained relatively constant over a wide range of Ea/Ees.

Effects of Myocardial Oxygen Consumption and Energy Conversion Efficiency

Myocardial oxygen consumption and energy conversion efficiency obtained in the three groups are presented below. Myocardial oxygen consumption analysis was available from 8 patients in the dobutamine group. For the dobutamine group, myocardial oxygen consumption increased from 5.52 ± 0.86 to 7.56 ± 1.52 J/beat

($p<0.05$), external work increased from 0.77 $\pm$ 0.08 to 0.95 $\pm$ 0.11 J/beat ($p<0.05$), and pressure-volume area decreased from 2.71 $\pm$ 0.44 to 2.27 $\pm$ 0.39 J/beat ($p<0.05$). As a result of these findings, the cardiac work efficiency increased from 31.7 $\pm$ 4.0 to 44.7 $\pm$ 3.6% ($p<0.001$), the ratio of PVA to MVO_2 decreased from 50.7 $\pm$ 4.7 to 30.7 $\pm$ 1.6% ($p<0.001$), and the mechanical efficiency decreased slightly from 15.9 $\pm$ 1.8 to 13.7 $\pm$ 1.7% ($p<0.001$).

Myocardial oxygen consumption analysis was available from 8 patients in MS-857 group. For the MS-857 group, changes in myocardial oxygen consumption, external work, and pressure-volume area were minimal and not statistically significant. Nevertheless, the energy conversion efficiencies showed significant changes. The cardiac work efficiency increased from 36.7 $\pm$ 4.3 to 48.3 $\pm$ 4.4% ($p<0.001$), the ratio of PVA to MVO_2 decreased from 51.4 $\pm$ 5.4 to 35.2 $\pm$ 2.5% ($p<0.01$), and the mechanical efficiency did not change significantly.

Myocardial oxygen consumption analysis was available from all 7 patients in OPC-18790 group. For the OPC-18790 group, myocardial oxygen consumption decreased from 4.97 $\pm$ 0.86 to 4.02 $\pm$ 0.73 J/beat ($p<0.05$), external work did not change significantly, and pressure-volume area decreased from 1.89 $\pm$ 0.22 to 1.24 $\pm$ 0.21 J/beat ($p<0.05$). As a result of these findings, the cardiac work efficiency increased from 41.8 $\pm$ 4.1 to 60.7 $\pm$ 6.1% ($p<0.01$), the ratio of PVA to MVO_2 did not change significantly, and the mechanical efficiency increased slightly from 17.9 $\pm$ 2.9 to 20.0 $\pm$ 3.2% ($p<0.01$).

Comparison of Effects Among Inotropic Agents

Comparative changes in indices of hemodynamics and energetics among the three group are shown in Figure 6 and 7. All the inotropic agents had a potent inotropic effect but they had different effects on heart rate, loading conditions and myocardial energetics.

Dobutamine increased heart rate while the other two inotropic agents did not. Dobutamine increased blood pressure while the other agents decreased it. All drugs significantly decreased left ventricular end diastolic volume. OPC-18790 remarkably decreased it, and had the most potent preload reducing effect.

All drugs increased Ees and improved Ea/Ees, the index of ventriculoarterial coupling. The drugs differed in their effect on Ea. Dobutamine tended to increase Ea, while OPC-18790 tended to decrease it. MS-857 significantly decreased Ea. MS-857 and OPC-18790 acted mainly as phosphodiesterase inhibitors, but there were differences in their effects on preload and afterload. MS-857 had a larger afterload reducing effect, whereas, OPC-18790 had a larger preload reducing effect than the others.

The effect on myocardial oxygen consumption also showed some difference among OPC-18790 and the other two drugs. Dobutamine and MS-857 increased the coronary blood flow, but had no significant effect on arteriocoronary oxygen content difference. On the other hand, OPC-18790 decreased arteriocoronary oxygen content difference, and did not increase coronary blood flow. Only OPC-18790 decreased myocardial oxygen consumption.

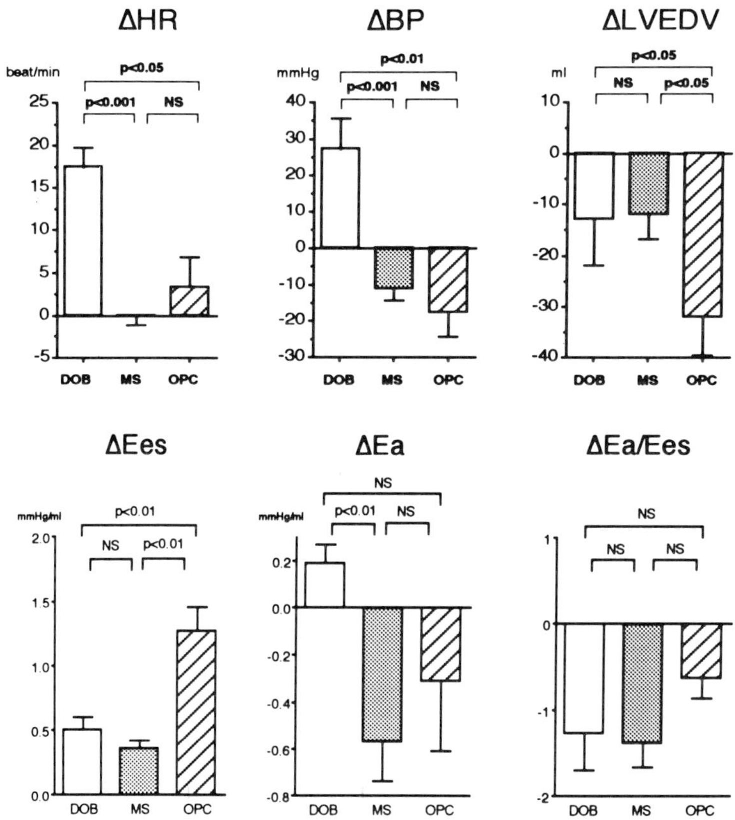

Fig. 6. Comparison of effects of inotropic agents on hemodynamics.

These different effects of the inotropic agents consequently resulted in the difference in energy conversion efficiencies. All drugs increased the cardiac work efficiency

EW/PVA by improving ventriculoarterial coupling. There was no significant difference in the first and second stage of energy conversion efficiencies among the

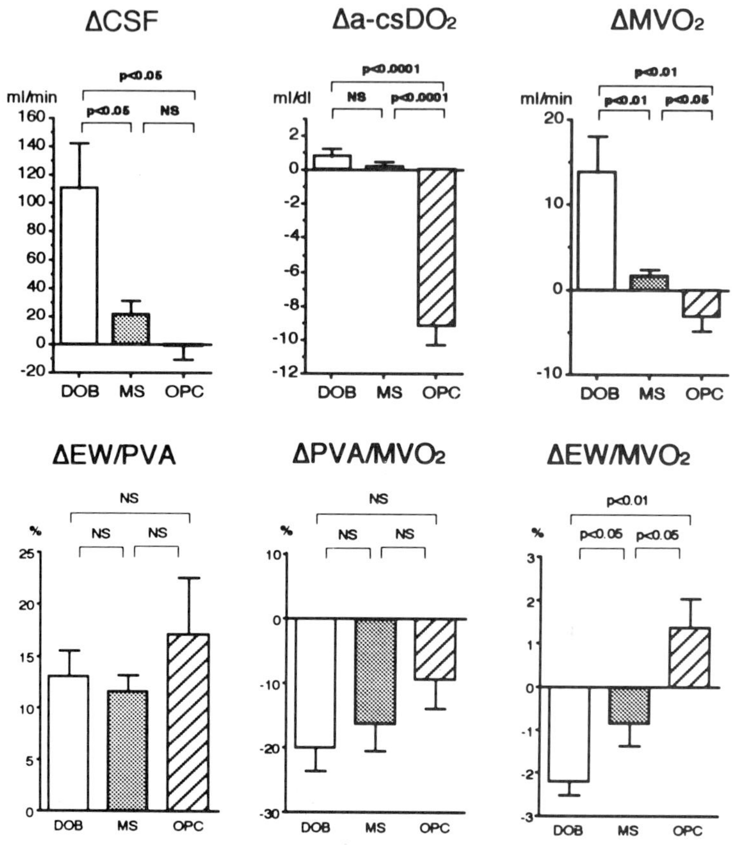

Fig. 7. Comparison of effects of inotropic agents on myocardial energetics.

three drugs. However, the effect on mechanical efficiency was significantly different among these three drugs. Dobutamine decreased mechanical efficiency. MS-857 slightly decreased mechanical efficiency but the change was not statistically significant. On the other hand, OPC-18790 increased mechanical efficiency and consequently had a beneficial effect on myocardial energetics.

DISCUSSION

This study has shown that a mismatching of the baseline ventriculoarterial coupling occurs with a large effective arterial elastance and small end-systolic elastance in patients with DCM. Patients with DCM have large left ventricles with depressed contractility. In order to maintain the peripheral circulation, the heart must generate sufficient blood pressure and flow to overcome the peripheral resistance. The left ventricle of a failing heart must enlarge (i.e., end-diastolic volume to V_o, where V_o is the volume axis intercept of end-systolic pressure-volume line) in order to maintain an adequate blood pressure. Therefore, the enlarged ventricles seen in idiopathic dilated cardiomyopathy patients could be considered to be a compensatory response to depressed contractility. The pathologic ventricular enlargement seemed to be a consequence of compensation and not to be a limiting factor to depressed hemodynamics and energetics.

After the administration of the above inotropic agents, Ea/Ees became smaller but remained greater than 1 in most of the patients. Under normal physiologic conditions, the ventricular and arterial properties of control patients are matched to maximize efficiency since Ea/Ees was in the range of 0.5 to 1. This study has demonstrated that the coupling conditions were far less than optimal and that cardiovascular performance is severely depressed both mechanically and energetically in patients with DCM.

Mechanical efficiency, defined as the ratio of external work to myocardial oxygen consumption, is the conversion efficiency from metabolic energy to the hydraulic energy that the left ventricle generates to the vascular system. This study showed that the transfer efficiency from myocardial oxygen consumption to pressure-volume area varied over a wide range, and had a close positive correlation with Ea/Ees. It is well known that the pressure-volume area has a close linear correlation with myocardial oxygen consumption, and can be expressed as the equation: $MVO_2 = A \bullet PVA + B$, where A and B are constants. The relationship between pressure-volume area and myocardial oxygen consumption is known to shift upward when contractility is enhanced and end-systolic elastance increases. Thus, the transfer efficiency from myocardial oxygen consumption to pressure-volume area changes according to changes in contractility or loading conditions, even if contractile efficiency is constant. This former efficiency will decrease when the ratio of effective arterial elastance to end-systolic elastance decreases toward 1 as was observed in this study.

The transfer efficiency from pressure-volume area to external work also varied over a wide range, and had a close curvilinear correlation with Ea/Ees. In this study, the experimentally derived regression equation closely resembled the equation predicted theoretically (28). This implies that the work efficiency can be evaluated only by measuring the ventriculoarterial coupling.

The overall efficiency from myocardial oxygen consumption to external work, the mechanical efficiency, remained constant over the range of Ea/Ees obtained in this study. This indicated that the mechanical efficiency would not be affected by the change of Ea/Ees in compensated, stable conditions of chronic heart failure but depended on the interaction of the first and the second steps of energy transfer. Different inotropic agents had slightly different effects on ventriculo-arterial coupling and the mechanical efficiency. Each drug improved the ventriculo-arterial coupling. The observed difference among the drugs was mainly in their effects on the preload and afterload. The pure phosphodiesterase inhibitor MS-857 had a vasodilating effect and led to a reduction in Ea. OPC-18790, which also acted mainly as a phosphodiesterase inhibitor, tended to decrease Ea, but the effect was smaller than MS-857. Dobutamine had an opposite effect on Ea. The most notable difference among these drugs was their effect on the mechanical efficiency. OPC-18790 increased the mechanical efficiency. MS-857 did not alter the efficiency. Dobutamine slightly decreased it. We believe that these differences are mainly due to their differences in provoking a vasodilating effect. Vasodilating action will decrease PVA and will tend to offset the increase in PVA dependent myocardial oxygen consumption which is induced by an increase in contractility. The preload reducing effect of OPC-18790 was the greatest of the three drugs, it decreased ventricular size, and might reduce the myocardial oxygen consumption. Moreover, external work would remain unchanged because of the inotropic action of OPC 18790, and consequently the mechanical efficiency might increase. Pure inotropic agents such as dobutamine require increased oxygen use in order to improve cardiac performance;however, they can have an oxygen wasting effect on the depressed heart of DCM. This adverse effect can be considered to be one of the difficulties in using inotropic agents for dilated, depressed hearts. In contrast, the fact that vasodilator therapy is rational therapy for improving hemodynamic performance without increasing myocardial oxygen requirements was confirmed in this pressure-volume analysis.

CONCLUSIONS

The coupling condition in patients with idiopathic dilated cardiomyopathy was far from the so-called "optimal" coupling. Moreover, the cardiovascular performance of these patients was severely depressed, both mechanically and energetically even in the mild to moderate cases. Dobutamine, MS-857, and OPC-18790 similarly improved the relationship between contractility and vascular load and increased the work efficiency, but not enough to optimize the coupling condition. These inotropic agents had minimal but different effects on mechanical efficiency. OPC-18790 increased contractility, decreased myocardial oxygen consumption, and consequently improved mechanical efficiency. OPC-18790 may be beneficial in the treatment of failing hearts by improving hemodynamics without increasing myocardial oxygen demand.

ACKNOWLEDGEMENTS

We thank Drs. Takashi Miyahara, Yasuhiro Kodama, Tatsuyuki Matsunami, Makoto Watanabe, Mitsunori Iwase, Masahiko Ukai, Ryozo Kato, Kozo Nagata, Hirofumi Kanda, Yasuhito Nishinaka for their cooperation and Sadako Ichihashi for preparing the manuscript.

REFERENCES

1. Tanganelli P, Lenarda A, Bianciardi G, et al. Correlation between histomorphometric findings on endomyocardial biopsy and clinical findings in idiopathic dilated cardiomyopathy. Am J Cardiol 64:504-506, 1989.

2. Figulla, H.R., Rahlf, G., Nieger, M., Luig, H,, Kreuzer, H. Spontaneous hemodynamic improvement or stabilization and associated biopsy findings in patients with congestive cardiomyopathy. Circulation 71:1095-1104 1985.

3. Douglas, P.S., Morrow, R., Ioli, A., Reichek, N. Left ventricular shape, afterload and survival in idiopathic dilated cardiomyopathy. J Am Coll Cardiol 13:311-315, 1989.

4. Juilliere, Y., Danchin, N., Briancon, S., et al. Dilated cardiomyopathy: long-term follow-up and predictors of survival. Int J Cardiol 21:269-277, 1988.

5. Katz, A.M. Potential deleterious effects of inotropic agents in the therapy of chronic heart failure. Circulation 73(suppl III):III-184-190, 1986.

6. Hasenfuss, G., Holubarsch, C., Heiss, H.W., et al. Myocardial energetics in patients with dilated cardiomyopathy. Circulation 80:51-64, 1989.

7. Caforio, A.L.P., Stewart, J.T., McKenna, W.T. Idiopathic dilated cardiomyopathy. Rational treatment awaits better understanding of pathogenesis. Br Med J 300:890-891, 1990.

8. Shah, P.K., Amin, D.K., Hulse, S., Shellock, F., Swan, H.J.C. Inotropic therapy for refractory congestive heart failure with oral fenoximone (MDL-17, 043): Poor long-term results despite early hemodynamic and clinical improvement. Circulation 71:326-331, 1985.

9. Beber, K.T., Janicki, J.S., Jain, M.C. Enoximone(MDL 17,043) for stable, chronic heart failure secondary to ischemic or idiopathic cardiomyopathy. Am J Cardiol 58:589-595, 1986.

10. Webster, M.W.I., Sarpe, D.N. Adverse effects associated with newer inotropic agents. Med Toxicol 1:335-342, 1986.

11. Packer, M., Medina, N., Yushak, M. Hemodynamic and clinical limitations of long-term inotropic therapy with Amrinone in patients with severe chronic heart failure. Circulation 70:1038-1047, 1984.

12. Suga, H., Igarashi, Y., Yamada, O., Goto, Y. Mechanical efficiency of the left ventricle as a function of preload, afterload, and contractility. Heart Vessels 1:3-8, 1985.

13. Suga, H., Hayashi, T., Shirahata, M. Ventricular systolic pressure-volume area as a predictor of cardiac oxygen consumption. Am J Physiol 240:H39-H44, 1981.

14. Suga, H., Hisano, R., Goto, Y., Yamada, O., Igarashi, Y. Effect of positive inotropic agents on the relation between oxygen consumption and pressure volume area in canine left ventricle. Circ Res 53:306-318, 1983.

15. Suga, H., Yasumura, Y., Nozawa, T., Futaki, S., Igarashi, Y., Goto, Y. Prospective prediction of O2 consumption from pressure-volume area in dog hearts. Am J Physiol 252:H1258-1264, 1987.

16. Suga, H. Total mechanical energy of a ventricle model and cardiac oxygen consumption. Am J Physiol 236:H498-H505, 1979.

17. Suga,H., Sagawa, K., Shoukas AA. Load independence of the instantaneous pressure-volume ratio of the canine left ventricle and effects of epinephrine and heart rate on the ratio. Circ Res 32:314-322, 1973.

18. Sunagawa, K., Maughan, W.L., Burkhoff, D., Sagawa, K. Left ventricular interaction with arterial load studied in isolated canine ventricle. Am J Physiol 245:H773-H780, 1983.

19. Sunagawa, K., Maughan, W.L., Sagawa, K. Optimal arterial resistance for the maximal stroke work studied in isolated canine left ventricle. Circ Res 56:586-595, 1985.

20. Report of the WHO/ISFC Task Force on the Definition and Classification of Cardiomyopathies. Br Heart J 44:672-673, 1980.

21. Goodwin, J.F. The frontiers of cardiomyopathy. Br Heart J 48:1-18, 1982.

22. Baan, J., van der Velde, E.T., de Bruin, H.G., et al. Continuous measurement of left ventricular volume in animals and humans by conductance catheter. Circulation 70:812-823, 1984.

23. Dodge, H.T., Sandler, H., Baxley, W.A., Hawley, R.R. Usefulness and limitations of radiographic methods for determining left ventricular volume. Am J Cardiol 18:10-24, 1966.

24. Ishihara, H., Yokota, M., Sobue, T., Saito, H. Relation between ventriculoarterial coupling and myocardial energetics in patients with idiopathic cardiomyopathy. J Am Coll Cardiol 23:406-416, 1994.

25. Ganz, W., Tamura, K., Marcus, H.S., Donoso, R., Yoshida, S., Swan, H.J.C. Measurement of coronary sinus blood flow by continuous thermodilution in man. Circulation 4:181-195, 1971.

26. Maruyama, M., Kamiya, J., Kaiho, T., Kobari, T. Cardiovascular properties of MS-857, a new and potent cardiotonic agent, on normal and failing hearts. J Cardiovasc Pharmacol 12:579-586, 1988.

27. Kamiya, J., Maruyama, M., Kaiho, T., Kobari, T. Characterization of MS-857, a new and potent cardiotonic agent, in isolated mammalian cardiac muscles. J Cardiovasc Pharmacol 14:790-797, 1989.

28. Nozawa, T., Yasumura, Y., Futaki, S., Tanaka, N., Uenishi, M., Suga, H. Efficiency of energy transfer from pressure-volume area to external mechanical work increases with contractile state and decreases with afterload in the left ventricle of the anesthetized closed-chest dog. Circulation 77:1116-1124, 1988.

Index